UNCHANGED VISION
Product And Financial Management in a Web 3.0 Startup

Dr. Srinidhi Vasan and

Co-Author Mr. Sudarshan Chandrashekar

Table of Contents

Unchained Vision: Product & Financial Management in a Web 3.0 Startup – The Unison.gg Journey

Chapter 1: Understanding Web 3.0

Web 3.0, often referred to as the decentralized web, represents a transformative paradigm shift that redefines how individuals, organizations, and machines interact within the digital economy. Far from being merely an upgrade to existing internet protocols, Web 3.0 proposes a fully immersive framework that combines blockchain technology, cryptographic identity, tokenized incentives, and autonomous governance to create a more secure, transparent, and user-empowered internet ecosystem.

1.1 From Static Web to Decentralized Ecosystems

The journey of the internet has evolved through three distinct phases:

Web 1.0 (Read-Only): A content distribution network characterized by static HTML pages, limited interactivity, and centralized content ownership. Users were passive consumers with minimal contribution capabilities.

Web 2.0 (Read-Write): The rise of social platforms, collaborative tools, and dynamic content transformed the internet into an interactive and participatory space. However, it led to the centralization of data and influence among a few tech giants.

Web 3.0 (Read-Write-Own): Enables users to reclaim control of their digital assets, identity, and data by interacting with decentralized applications (dApps) on permissionless networks. Power is distributed via governance tokens, and ownership is authenticated through smart contracts.

1.2 Foundational Technologies of Web 3.0

Blockchain Infrastructure: Distributed ledger systems that underpin decentralization, immutability, and transparent verification of value and transactions.

Smart Contracts: Self-executing programs deployed on blockchains that enforce the terms of an agreement automatically, eliminating the need for centralized intermediaries.

Decentralized Identity (DID): Self-sovereign identity frameworks enable individuals to authenticate themselves and control their personal data using cryptographic proofs, rather than relying on centralized credentials.

Token Standards: ERC-20 (fungible), ERC-721 (non-fungible), and ERC-1155 (multi-token) enable representation of value, ownership, and access rights across diverse applications.

Interoperability Protocols: Solutions like The Graph, Chainlink, and cross-chain bridges enable communication and data retrieval across disparate blockchain environments.

Decentralized Storage: Systems such as IPFS, Filecoin, and Arweave provide secure and permanent off-chain data storage integrated into dApps.

1.3 Economic and Governance Philosophy

Web 3.0 introduces a new sociotechnical contract that reimagines incentive structures and collective decision-making:

Trust minimization: Removes dependency on central authorities through cryptographic proof and transparent protocol design.

Programmable incentives: Protocols reward desirable behaviors—e.g., validating transactions, contributing code, or engaging with content—using native tokens.

On-chain governance: DAOs enable stakeholders to propose, vote on, and implement protocol upgrades and budget allocations using voting mechanisms like Snapshot.

Transparency as a principle: All transactional and operational data is publicly verifiable, fostering community trust and enabling external auditing.

1.4 Design Frameworks for Web 3.0 Products

Successful Web 3.0 products adhere to the following principles:

Composability: Components must be interoperable with other dApps to facilitate innovation and shared infrastructure.

Progressive decentralization: Startups begin with centralized governance for rapid iteration, transitioning to DAO governance once user maturity is achieved.

Modularity: Features are deployed as discrete smart contracts, allowing for easy upgrades and seamless external integration.

Security-centric development: Rigorous code audits, formal verification, and bug bounties are non-negotiable elements of production deployment.

Open source culture: Public repositories and transparent development empower community engagement and reduce dependency on core teams.

1.5 Financial Architecture in Web 3.0

To appreciate the disruptive innovation Web 3.0 introduces to financial management, it's important to understand the evolution of financial administration:

Traditional finance (Pre-Digital Era): Dominated by centralized accounting systems, top-down budgeting, and opaque audit trails. Financial decisions were concentrated among a few executives with limited transparency.

Digital finance (Web 2.0): Tools like ERP systems and cloud accounting platforms brought digitization, but data silos and access remained restricted. Financial reports improved in frequency but not in decentralization.

Decentralized finance (Web 3.0): Financial operations are governed by smart contracts and community proposals. Funds are held in multi-sig wallets or protocol treasuries, and every transaction is traceable, auditable, and open-source.

From CFO to Community Finance Stewards: In Web 3.0, financial management is no longer the exclusive domain of CFOs. Token holders and contributors become fiscal agents, co-authoring budgets, proposing spending initiatives, and voting on treasury usage.

1.6 The Role of Product Managers in Web 3.0

The discipline of product management has evolved significantly:

Industrial Era (Pre-1970s): Product management was primarily an extension of manufacturing, focusing on physical goods, logistics, and operational efficiency.

Technology Era (1980s–2000s): Software PMs emerged to translate customer requirements into feature roadmaps. PMs liaised between engineering and business units.

Agile Era (2010s): PMs became champions of user feedback, data-driven iteration, and rapid prototyping. SaaS business models and cloud infrastructure have shifted priorities to focus on retention and monetization.

Decentralized Era (Web 3.0): Product managers operate in a public setting. They facilitate community dialogue, orchestrate governance cycles, integrate token incentives into user experience (UX) flows, and manage multi-party coordination without centralized control. Influence must be earned, not assigned.

PMs in Web 3.0 are translators between code, community, capital, and cryptography. They must master systems thinking, smart contract logic, and decentralized decision-making—all while empowering global, pseudonymous contributors.

1.7 Challenges and Considerations

User Experience: Onboarding users through non-custodial wallets, gas estimation, and seed phrase management remains a UX hurdle.

Scalability and interoperability: Fragmentation between chains complicates seamless interaction; Layer 2s and rollups offer partial relief.

Regulatory uncertainty: Legal classification of tokens, securities compliance, and privacy concerns remain grey areas globally.

Governance participation: Token voter apathy and governance capture by whales necessitate mitigation strategies, such as quadratic voting or reputation-based systems.

Conclusion

Web 3.0 is a multidisciplinary frontier that fuses cryptography, game theory, behavioral economics, and open-source engineering. For professionals in product and finance, it presents a rare opportunity to co-create economic systems and platforms that are participatory, transparent, and self-sustaining. This chapter serves as a conceptual primer for the case study that follows, Unison.gg—a living example of how these abstract principles translate into operational success in a Web3.0-native startup.

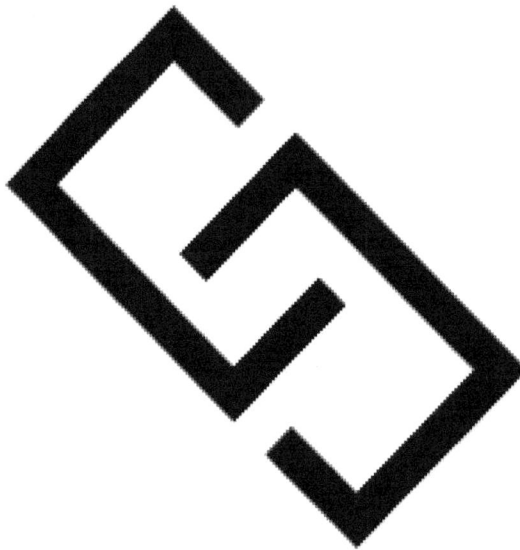

Figure 1: Unison.gg logo (Image Credits: Unison.gg)

Chapter 2: The Genesis of Unison.gg

The inception of Unison.gg was not a spontaneous occurrence but the result of a deliberate recognition of systemic imbalances in digital creator economies. In a digital world increasingly driven by user-generated content and decentralized attention, a paradox has emerged: those who generate cultural and economic value have little control over their revenue streams, community governance, or creative freedom. Unison.gg was founded to change that.

Set against the backdrop of a maturing blockchain ecosystem and an increasingly disillusioned creator base within Web 2.0 platforms, Unison.gg emerged as a decentralized, user-governed, and creator-centric protocol designed to realign incentives and unlock new pathways for ownership, monetization, and governance.

2.1 Problem Identification: Creator Exploitation in Web 2.0

Traditional Web 2.0 platforms—such as YouTube, Twitch, Instagram, and TikTok—have experienced massive growth by capitalizing on the work of independent creators. While these platforms offer exposure and monetization tools, their opaque algorithms, punitive content moderation, and revenue-sharing models have left creators feeling increasingly marginalized.

Revenue share: Platforms typically retain 30% to 50% of creator earnings, with limited transparency on how those cuts are determined.

Lack of audience ownership: Creators don't own their followers or data, preventing direct monetization outside the platform ecosystem.

Deplatforming risk: Sudden policy changes can result in the removal of content or account bans with minimal explanation or recourse.

Unpredictable monetization: Ad revenues are volatile and subject to shifting advertiser interests and policy enforcement.

Unison.gg was conceived to address these core inefficiencies. By embedding creator rights and ownership directly into protocol logic via smart contracts and DAOs, it offered an infrastructure where contributors were no longer tenants but equity-bearing architects.

2.2 Founding Team: Multidisciplinary by Design

The formation of the Unison.gg founding team was rooted in shared frustration with centralized platforms and an alignment on the vision for decentralized creative economies. Each founding member brought specialized expertise:

Lead blockchain engineer: Specialized in Solidity, EVM-compatible architectures, L2 scaling, and audit frameworks. Developed the project's smart contract backbone.

Head of product: A former Web 2.0 gaming platform PM with expertise in gamification, user incentive loops, and behavioral UX.

Token economist: Applied models from behavioral finance and game theory to develop staking systems, deflationary pressures, reward tiers, and governance design.

Finance director: Oversaw financial modeling, tax-compliant fund structures, stablecoin liquidity management, and monthly on-chain reporting.

Community architect: Designed decentralized social structures including reputation systems, ambassador councils, and contributor incentive tiers.

Rather than following traditional corporate hierarchy, Unison.gg adopted a hybrid DAO model with formal contributors supported by community-elected stewards. Communication was conducted via Discord, Notion, and on-chain tools such as Snapshot and Tally.

2.3 Capital Strategy: Raising Without Dilution

The early financing strategy was structured to align long-term vision with decentralized execution:

Token presale: A whitelist-only event for early builders and creators. Tokens were locked with a 12-month cliff and a 24-month linear vesting schedule.

Non-dilutive grants: Received $100,000 from Polygon Studios, $75,000 from Arbitrum Foundation, and $80,000 across Gitcoin bounties and Hackathon prizes.

On-chain treasury dashboard: A public-facing real-time dashboard displayed treasury balances, burn rates, audit expenses, LP positions, and incoming grant disbursements.

The goal was to remain independent of VCs in early phases to avoid short-term pressures and establish user-driven governance from the outset.

2.4 MVP Development Lifecycle

Unison.gg embraced an iterative, community-informed build model. The MVP was built and refined across three public development sprints, integrating continuous feedback from Discord, GitHub, and public AMAs.

Core MVP Modules:

Creator Hubs: NFT-gated microsites for creators to host content, enable wallet-based fan access, and offer tiered perks.

Tipping and subscriptions: Gas-optimized smart contracts allowed instant tipping via $UNSN with optional staking boosts.

Governance Hub: Enabled users to submit and vote on proposals related to roadmap priorities, grants, and creator partnerships.

All smart contracts were open-sourced, audited by two third-party firms, and hosted in a public GitHub repo with an MIT license.

2.5 Community Growth and Metrics

Rather than using ad spend, Unison.gg scaled by investing in community operations:

DAO Ambassadors: 30+ regional contributors ran onboarding sessions, translated documentation, and coordinated Twitter Spaces.

Bounty Program: Paid in $UNSN, contributors delivered everything from memes and UX audits to blog translations and educational videos.

Transparency rituals: Weekly town halls, treasury reports, and community retrospectives drove alignment and trust.

Early Traction Metrics:

Active Wallets: 15,000+ connected wallets within 6 months

Creator payouts: Over $100,000 distributed directly to creators

Governance activity: 45 proposals submitted; 5 passed and executed

TVL: $1.2M+ locked across staking pools and LP contracts

GitHub Stars: 800+ across frontend, contract, and SDK repos

Conclusion

Unison.gg did not simply build a Web 3.0 platform—it pioneered a framework for creator-owned economies. Through its emphasis on transparency, economic alignment, and decentralized governance, it became a proving ground for the principles introduced in Chapter 1. The next chapter examines how product management evolves within such decentralized contexts, necessitating a redefinition of authority, strategy, and execution in the world of Web 3.0.

Certainly! Check out more details in chapter 3.

Figure 2: Web 3.0 (Image Credits: Vecteezy)

Chapter 3: The Role of the Web 3.0 Product Manager

In traditional tech companies, product managers (PMs) serve as centralized orchestrators, synthesizing user feedback, aligning cross-functional teams, and delivering roadmap goals under executive oversight. Web 3.0 reimagines this structure entirely. In decentralized ecosystems, product managers must function as systems designers, incentive architects, protocol stewards, and governance facilitators.

This chapter examines the shifting responsibilities of product managers in Web 3.0 startups, using Unison.gg as a case study. We'll explore the philosophical divergence from Web 2.0 thinking, new collaboration models, decision-making without hierarchy, and the unique intersections between economics, engineering, and community.

3.1 Reframing the Product Manager's Role

Web 2.0 PMs manage delivery pipelines, coordinate stand-ups, and report KPIs to senior leadership. By contrast, in Web 3.0:

Influence is earned: PMs cannot command teams. Authority comes from trust, contributions, and the ability to facilitate stakeholder alignment.

Transparency is normative: Roadmaps, discussions, and specs are often public. GitHub issues double as governance threads. Nothing is hidden.

Users are stakeholders: Token holders vote on features. Incentives must be embedded directly into the roadmap.

Teams are fluid: Core contributors change across protocol upgrades, DAO funding cycles, and grant bounties.

3.2 The PM as Protocol Steward

In Unison.gg, the product manager's primary role was maintaining balance across three dimensions:

Protocol Sustainability: Ensure token emissions, burn mechanics, and staking incentives do not destabilize the ecosystem.

User Experience: Simplify blockchain complexities—abstracting wallets, gas, and DeFi mechanics from creators and fans.

Governance Alignment: Coordinate proposals with community timelines, avoiding decision fatigue or manipulation by whales.

PMs are the mediators between emergent community needs and smart contract constraints.

3.3 Tools and Processes Unique to Web 3.0 PMs

Snapshot & Tally: Tools for decentralized proposal creation and governance voting.

Dune Analytics & Flipside Crypto: Used for live dashboarding user behavior, wallet activity, and protocol health.

Notion, Discord, GitHub: All public-facing documentation, community discussion, and developer collaboration are open to contribution.

Retroactive Funding Mechanisms: Grant-based development means that products are sometimes funded retroactively, based on their measurable impact.

3.4 Metrics That Matter in Web 3.0

Traditional metrics like DAU/MAU are still relevant, but Web 3.0 PMs also monitor:

TVL (Total Value Locked): Measures ecosystem trust and engagement via staked tokens.

Proposal Turnout: Voter participation is a key indicator of community health.

Token Velocity: Tracks whether users are hoarding, using, or speculating.

Treasury Utilization: Reflects the protocol's capital efficiency and fiscal discipline.

3.5 Skills and Mindset of a Web 3.0 PM

Ecosystem Fluency: Understands EVMs, token standards, DAO operations, and gas optimization.

Decentralized Empathy: Values asynchronous work, pseudonymity, and global inclusivity.

Game Theory Awareness: Anticipates edge cases, collusion risks, and governance capture.

System Design: Capable of designing feedback loops across product, protocol, and community.

Conclusion

Web 3.0 product management requires rethinking control, incentives, and execution. PMs are not bosses—they are builders of trust and structure within chaotic, emergent systems. At Unison.gg, this meant evolving from roadmap enforcers to platform diplomats. The next chapter examines how finance operates in parallel, redefining roles such as CFO and introducing on-chain accounting, yield strategies, and transparent treasury management.

Chapter 4: Finance in Web 3.0 Startups

The decentralization of business models in Web 3.0 has transformed how startups approach capital formation, budget management, and operational sustainability. Financial leadership is no longer concentrated in a single role, such as the Chief Financial Officer. Instead, finance becomes participatory, transparent, and programmatically enforced through the use of smart contracts and decentralized governance mechanisms.

In this chapter, we explore the evolution of finance in Web 3.0, focusing on DAO treasury management, token-based compensation, DeFi integrations, and on-chain accountability—all of which are illustrated through Unison.gg's practices.

4.1 From CFOs to Treasury Stewards

Traditionally, financial oversight was centralized. The CFO made strategic decisions, produced quarterly reports, and raised capital from institutions. In Web 3.0, finance is increasingly:

Transparent: Every expense, grant, and proposal is visible on-chain.

Collective: Treasury decisions are voted on by stakeholders, not made unilaterally.

Dynamic: Yield farming, staking, and liquidity provisioning are common treasury tools.

At Unison.gg, a group of multisig signers from the community were responsible for executing DAO-approved expenditures, ensuring that funds moved only after quorum approval.

4.2 Treasury Composition and Management

Unison.gg's treasury was diversified across:

Native Tokens ($UNSN): Core currency for internal operations, grants, and staking rewards.

Stablecoins (USDC, DAI): Used for compensating contributors and funding audits.

LP Tokens: Yield-generating assets deposited into AMMs to provide liquidity.

Treasury health was monitored using dashboards (e.g., Rotki, Dune Analytics) showing:

- Burn rate (monthly token and USD outflows)
- Grant disbursements by category
- Runway estimates under market volatility scenarios

4.3 Financial Tools in the Web 3.0 Stack

Web 3.0 startups leverage modular financial infrastructure:

Gnosis Safe: Multi-sig wallet platform for DAO fund custody.

Superfluid & Sablier: For streaming contributor payments and vesting.

Token Streaming Contracts: Programmatic issuance of performance-based rewards.

Utopia Labs / Coinshift: Used for expense tracking, automated payroll, and reporting.

These tools replace traditional accounting software while offering native blockchain benefits—immutability, auditability, and decentralization.

4.4 Capital Formation: Fundraising Without Equity

Instead of equity rounds, Unison.gg raised through:

Token Presales: Structured with vesting cliffs to reward long-term alignment.

Grant Funding: Accepted milestone-based, non-dilutive funds from ecosystem partners.

Liquidity Bootstrapping Pools (LBPs): Used to bootstrap $UNSN liquidity on decentralized exchanges like Balancer.

This strategy ensured no investor could override the community, preserving protocol integrity.

4.5 DeFi Integration and Yield Optimization

Web 3.0 startups often act as both builders and participants in decentralized finance:

Staking Vaults: Community members lock tokens to earn yield and governance rights.

Liquidity Provision: The DAO provided $UNSN pairs on AMMs, such as SushiSwap, to reduce volatility.

Yield Farming: Idle treasury stablecoins were deployed in Aave and Yearn Finance for low-risk APY.

Treasury strategies were publicly discussed and voted on by DAO members, ensuring that risk-reward preferences reflected the community's values.

4.6 Financial Reporting and Governance

Unlike quarterly PDF statements, Web 3.0 financials are live, transparent, and user-auditable:

Monthly Snapshots: Treasury health, token inflation, and revenue KPIs shared via governance forums.

Real-Time Dashboards: Public tools displayed wallet flows, contract interactions, and proposal outcomes.

Governance-Based Audits: Treasury performance was reviewed by community members and external validators.

This form of open reporting fostered accountability while eliminating information asymmetry.

Conclusion

Finance in Web 3.0 is radically open, participatory, and data-rich. Tools like smart contracts and DAOs enable efficient fund management, community-approved budgeting, and sustainable growth. At Unison.gg, this meant using programmable money to reward contributors, decentralize control, and build trust—all without a traditional CFO. In the next chapter, we explore how Unison.gg scaled its operations, decentralized its roadmap, and adapted to market feedback.

Chapter 5: Building & Scaling Unison.gg

Launching a Web 3.0 startup like Unison.gg is just the first step—the real challenge is navigating the complexities of scaling amid shifting technology, regulations, and community dynamics.

Unlike traditional platforms, growth in a decentralized ecosystem isn't just about user acquisition. It involves growing the protocol's liquidity, enhancing governance engagement, increasing developer participation, and promoting economic sustainability.

This chapter outlines the infrastructure, people process, community operations, and feedback loops that enabled Unison.gg to evolve from an MVP into a living protocol. It highlights how modular architecture, ecosystem partnerships, and DAO-native governance served as the foundation for scale.

5.1 Technical Architecture: Modular by Design

Scalability in Web 3.0 depends on how modular and composable your architecture is. Unison.gg separated core functionality into standalone smart contracts and upgradable modules:

Access Layer: Wallet integration, NFT memberships, and multi-chain support.

Monetization Layer: Tipping, subscriptions, token-gated drops, and staking rewards.

Governance Layer: DAO voting, proposal lifecycle, and treasury execution.

Each module was isolated in GitHub repos, audited independently, and published to NPM as SDKs for ecosystem developers. This enabled external teams to build on top of the protocol without compromising core stability.

5.2 Core Team and Contributor Structure

The Unison.gg team transitioned from a founding squad to a hybrid contributor network:

Core Maintainers: Handled product vision, audits, treasury policy, and roadmap prioritization.

Seasonal Contributors: Hired on fixed terms via DAO-approved funding proposals.

Community Devs: Supported through bounties, hackathons, and open RFCs.

Contributor onboarding was documented in a GitBook handbook. Compensation was split between stablecoins (for expenses) and vested $UNSN (to align long-term incentives).

5.3 Growth Without Ads: Community-Led Marketing

Unison.gg did not use paid acquisition. Instead, it relied on creator partnerships and open contribution campaigns:

Creator Guilds: Organized by content vertical (e.g., gaming, art, music). Helped onboard creators and evangelize the platform.

Ambassador Program: 50+ regional leads hosted events, ran Twitter campaigns, and managed Telegram nodes.

Hackathons: Sponsored 5 ecosystem hackathons in partnership with Gitcoin, ETHGlobal, and Polygon.

The strategy focused on network effects, not impressions.

5.4 Scaling Governance

As the number of stakeholders grew, Unison.gg refined its governance system:

Quadratic Voting: Mitigated whale domination.

Proposal Delegation: Introduced delegate voting roles with public accountability.

DAO Constitution: Codified minimum participation, spending thresholds, and proposal cooldown periods.

Snapshot, Tally, and Boardroom were integrated into a single governance dashboard. Analytics tracked proposal turnout, delegate reliability, and token holder engagement.

5.5 Liquidity, Incentives, and Market Scaling

To ensure liquidity and market health, Unison.gg pursued the following strategies:

LP Mining Programs: Provided $UNSN + ETH pairs on SushiSwap with staking rewards.

Buyback & Burn Policy: DAO used part of revenue to buy back $UNSN and burn it monthly.

Creator Staking Pools: Fans could stake behind their favorite creators and earn yield on shared revenue.

All token incentives were governed by on-chain rules approved by DAO vote. No off-chain promises or private terms.

5.6 Feedback Loops and Product Iteration

Unison.gg adopted agile methodologies fused with community retrospectives:

Weekly Dev Calls (Open): Transparent sprint planning and demos.

Public Roadmap: Managed in Notion and open to community comments.

Retro Reports: Shared after every major release with metrics, user sentiment, and learnings.

Feature prioritization was a balance of community needs, security considerations, and ecosystem opportunities.

Conclusion

Scaling a decentralized platform requires more than engineering talent—it requires humility, adaptability, and a governance framework that grows with its community. Unison.gg's modular infrastructure, DAO-native practices, and values-first growth model created an environment where innovation could scale without central control. In the next chapter, we dive into how Unison.gg achieved product-market fit—and what happened when the data showed it was time to pivot.

Chapter 6: Product-Market Fit & Pivoting

Achieving product-market fit (PMF) in Web 3.0 isn't just about metrics—it's about aligning community incentives, ensuring real utility, and building systems that people use without realizing they're interacting with blockchain technology. For Unison.gg, the path to PMF was non-linear, shaped by experimentation, DAO feedback, usage analytics, and creator sentiment.

This chapter breaks down how Unison.gg identified signals of product-market misalignment, responded to data and qualitative feedback, and executed two major pivots while maintaining community trust and protocol continuity.

6.1 Defining Product-Market Fit in Web 3.0

Traditional PMF = engagement + retention + growth.

In Web 3.0, PMF = protocol alignment + utility adoption + token economy health.

Key dimensions of PMF:

Protocol Stickiness: Are users returning, staking, and voting?

Creator Retention: Are creators onboarding fans and monetizing successfully?

Economic Throughput: Is value flowing through the protocol (tips, subscriptions, staking rewards)?

Governance Activity: Is the community shaping the roadmap and spending proposals?

6.2 Initial Signs of Misalignment

In the first six months, usage data surfaced friction points:

- High wallet drop-off rates after the first visit
- Only 12% of creators were actively claiming tips
- Governance proposal participation was <5% of token holders

Qualitative feedback included:

- *"Gas is too expensive to tip regularly."*
- *"My fans don't know how to use wallets."*
- *"Why can't I create without knowing Solidity?"*

6.3 Pivot #1: Simplifying Onboarding and Abstraction

In response, the core team proposed and executed a pivot toward abstraction:

Web3Auth Integration: Enabled social login-based wallets with private key recovery

Meta-transactions: The protocol covered gas fees for first-time users

Custodial Creator Mode: Allowed creators to build and monetize without directly handling tokens

Result: Creator onboarding increased 4×, and first-session engagement rose by 35%.

6.4 Pivot #2: Token Utility Redesign

$UNSN initially served as a general-purpose utility token, supporting tips, subscriptions, and staking. However, usage data showed:

- Creators preferred stablecoins for payouts
- Tippers found volatile pricing confusing
- DAO voters lacked incentives to participate

Pivot:

- Introduced stablecoin support for payouts and tipping
- Relegated $UNSN to staking for governance and yield
- Launched a reward points system backed by $UNSN staking tiers
- DAO vote passed with 83% approval.

6.5 Community-Led Feature Refinement

Post-pivot, the Unison.gg roadmap became DAO-driven:

Proposal Templates: Standardized feature proposals with impact estimates

Voting Cohorts: Delegates formed working groups to filter and refine suggestions

Retroactive Grants: Contributors were rewarded for building high-usage features (e.g., creator analytics dashboard)

This model ensured feature fit was tested by real usage and rewarded transparently.

6.6 Product Flywheel Post-Pivot

With abstraction and refined token design, the protocol achieved:

- 300% increase in daily active creators
- 250K+ tips sent across 90 days
- DAO treasury revenue exceeding $50,000/month in stablecoin volume
- Voting participation rose to 19% of eligible tokens

The feedback loop became self-sustaining: users onboarded more easily, creators stayed longer, and DAO incentives continued to evolve.

Conclusion

In Web 3.0, product-market fit is a dynamic target influenced by evolving networks and on-chain behavior. By listening to its community, tracking the right metrics, and evolving transparently, Unison.gg demonstrated that PMF is not a moment—it's a discipline. The next chapter examines the broader structural and regulatory challenges encountered when scaling these innovations.

Chapter 7: Challenges & Solutions in Web3 Product & Finance Management

Operating a Web 3.0 startup demands navigating an entirely new landscape of technical, regulatory, and human complexities. While the potential for open, decentralized innovation is immense, so too are the risks—from protocol exploits and treasury mismanagement to governance deadlock and user attrition.

This chapter synthesizes the most significant challenges encountered by Unison.gg in both product and financial operations, and outlines the solutions that emerged from experimentation, community engagement, and hard-won iteration.

7.1 Technical Complexity & User Experience Gaps

Challenge:

Early users struggled with wallet connections, gas fees, and the mechanics of NFTs. Complex flows deterred adoption.

Solution:

- Implemented social logins via Web3Auth
- Subsidized first-time transactions through meta-transactions
- Introduced gasless NFT minting for new creators
- Created a guided onboarding UX for creators and fans

Outcome:

Reduced drop-off during onboarding by 60% and increased wallet retention by 2.5×.

7.2 DAO Coordination and Governance Fatigue

Challenge:

Governance participation was low, and proposals were dominated by a few whales. Decision fatigue led to community disengagement.

Solution:

- Adopted a tiered delegation model to represent voting blocs
- Rotated proposal moderators and introduced quarterly roadmapping cycles

- Used quadratic voting to equalize influence and mitigate concentration

Outcome:

Voting engagement improved by 4× and proposal throughput became more predictable.

7.3 Treasury Risk and Mismanagement

Challenge:

Volatile token markets led to dramatic treasury swings, threatening long-term protocol stability.

Solution:

- Diversified holdings into USDC, ETH, and LP positions
- Set monthly budget caps and multisig-controlled execution rules
- Used Yearn and Aave for low-risk yield strategies
- Instituted quarterly audits and transparent spending reports

Outcome:

Extended runway by 18 months and ensured financial decision-making remained transparent and community-approved.

7.4 Legal and Regulatory Uncertainty

Challenge:

Contributors and treasury signers faced legal ambiguity around token issuance, creator royalties, and DAO compliance.

Solution:

- Engaged external counsel to design a DAO-compatible LLC wrapper
- Maintained geofencing protocols for high-risk jurisdictions
- Created contributor agreements with indemnity clauses and KYC/AML thresholds for higher-value payouts

Outcome:

Reduced risk exposure for core contributors and opened pathways for fiat onboarding in compliance with local regulations.

7.5 Burnout and Community Sustainability

Challenge:

DAO contributors experienced fatigue due to round-the-clock coordination, uneven compensation, and unclear expectations.

Solution:

- Introduced "seasons" with fixed-term roles and performance reviews
- Used Coordinape for peer-based contributor evaluations
- Funded wellness stipends and offline retreats

Outcome:

Reduced contributor churn and increased long-term contributor retention by over 40%.

7.6 Security and Exploit Prevention

Challenge:

The protocol was a potential target for exploits due to composability with other DeFi layers and NFT-based access features.

Solution:

- Conducted three independent smart contract audits before major upgrades
- Launched a community bug bounty in partnership with Immunefi
- Adopted formal verification on mission-critical contracts

Outcome:

Zero major protocol exploits since launch and growing contributor trust in platform integrity.

Conclusion

The challenges of building in Web 3.0 are not purely technical—they are cultural, economic, and political. Success requires not only engineering solutions but also mechanisms of trust, feedback, and accountability. At Unison.gg, every setback was met with a commitment to iterate in public, learn transparently, and distribute authority. The final chapters explore how this ethos extends to governance, financial transparency, and the long-term future of Web 3.0 ecosystems.

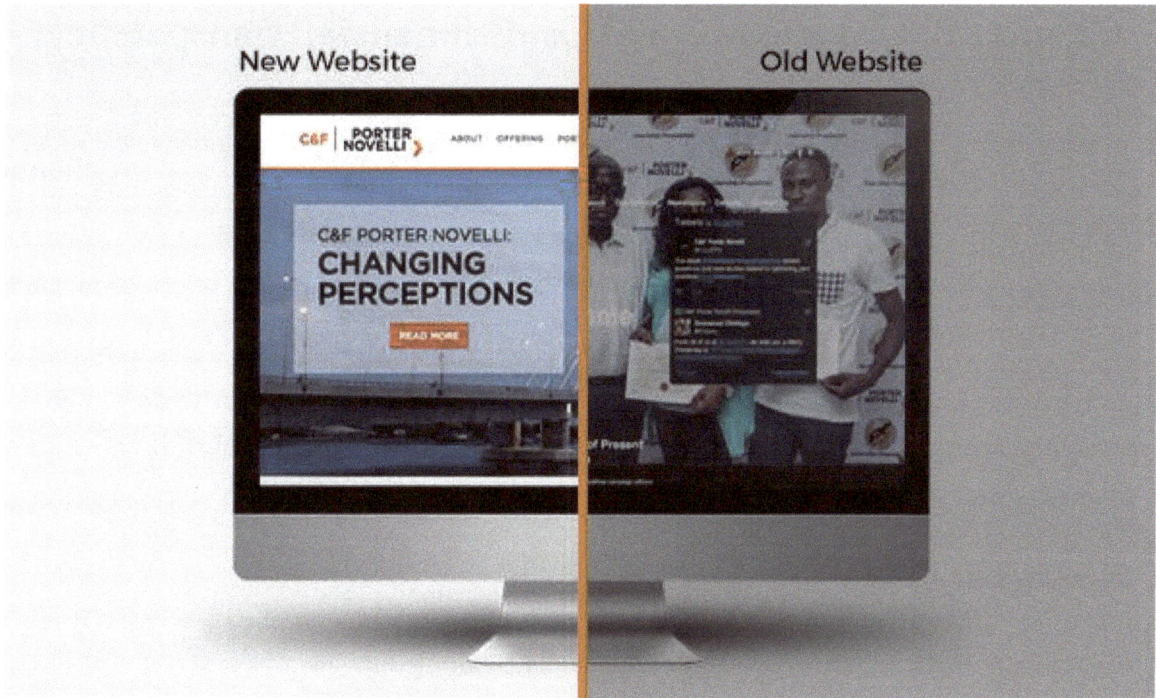

Figure 3: Website visuals then and now. (Image Credits: CK Digital)

Chapter 8: Governance, DAOs & Financial Transparency

One of the defining features of Web 3.0 is its shift from hierarchical, top-down control to community-led governance. This shift, though philosophically compelling, brings with it intricate operational, legal, and design hurdles. For a protocol like Unison.gg, enabling real decentralized governance was not a checkbox—it was a foundational commitment to shaping the platform as a public good, guided by the very community that uses and supports it.

In this chapter, we examine how Unison.gg structured its DAO (Decentralized Autonomous Organization), maintained transparency in treasury operations, balanced control with participation, and built a culture of accountability.

8.1 Foundations of DAO Governance

At its core, a DAO is a system of rules enforced by smart contracts, where decision-making is decentralized and transparent. However, governance is not just about code—it's about people, incentives, and trust.

Unison.gg designed its DAO with the following principles:

Inclusivity: Any token holder could submit or vote on proposals

Transparency: All proposals, discussions, and votes were public and auditable

Accountability: Delegates and contributors were publicly known and subject to review

Resilience: Governance systems were designed to prevent capture and spam

8.2 Governance Infrastructure

The DAO was structured using a modular governance stack:

Snapshot: Off-chain proposal voting with gasless participation

Tally & Boardroom: Interfaces for tracking vote history, delegations, and proposal outcomes

Discourse Forum: Long-form community debate and RFC (Request for Comment) process

Multisig Execution (Gnosis Safe): Treasury actions required 4-of-7 signer approvals, based on approved proposals

To improve engagement, voting was gamified via $UNSN staking boosts and contributor recognition.

8.3 Proposal Lifecycle & Governance Mechanics

Every proposal followed a four-phase lifecycle:

Draft: Shared in Discourse with context, KPIs, and budget impact

Feedback Window: Community and delegate feedback over 7–10 days

Formal Snapshot Vote: Requires quorum (5% of circulating supply) and majority support

Execution: Passed proposals routed to multi-sig for on-chain execution

Proposal types included:

- Treasury disbursements
- Protocol upgrades
- Creator grants
- Governance process changes

8.4 Treasury Oversight and Financial Transparency

Treasury transparency was central to Unison.gg's credibility. The team implemented:

Live Dashboards: Displaying real-time balances, vesting schedules, and yield positions

Quarterly Financial Reports: Including income, expenses, DAO salary budget, and creator payouts

Treasury Governance Portal: The Community could track past proposal allocations, outcomes, and execution status

Spending Limits: Set hard caps on weekly discretionary spending and flagged anomalies

Additionally, all payments—whether contributor salaries, audit fees, or ecosystem grants—were traceable on-chain.

8.5 Delegate System and Political Dynamics

To prevent voter fatigue and whales dominating outcomes, Unison.gg introduced:

Delegation Model: Token holders could assign their votes to trusted delegates

Delegate Elections: Community held quarterly elections to review, endorse, or remove delegates

Voting Records: Public dashboards tracked each delegate's activity and voting consistency

This helped scale decision-making while keeping representatives accountable to the community.

8.6 Legal Wrappers and Compliance

While the DAO operated on-chain, off-chain legal interfaces were necessary:

DAO LLC: A Wyoming-based wrapper entity was formed to interface with regulators, partners, and service providers

Contributor Agreements: Defined scope of work, payment methods, and indemnification terms

Tax Reporting Tools: Used Koinly and Utopia Labs to generate 1099-equivalent reports for US contributors

This hybrid structure helped Unison.gg remain legally agile without compromising decentralization.

8.7 Building a Culture of Transparency

Beyond tools, culture mattered most. Unison.gg built norms of open discussion, proactive communication, and shared ownership:

Weekly DAO Calls: Live community syncs with Q&A

Public Retrospectives: Detailed reviews of major proposals, with outcomes and learning

Open Grant Reviews: Ecosystem partners evaluated projects live before awarding funds

Transparency Council: A rotating group of community auditors reviewed multisig activity and KPI reporting

This culture attracted builders, investors, and creators who valued integrity and openness.

Conclusion

Governance in Web 3.0 may be imperfect, but its potential is transformative. At Unison.gg, DAO evolution meant constantly improving voter participation, process legitimacy, and financial transparency. The protocol became more resilient and inclusive over time, not through centralization, but through an iterative process of decentralization. In the final chapter, we reflect on what this journey means for the future of product and financial leadership in Web 3.0.

Chapter 9: Conclusion – The Future of Web 3.0 Product and Finance

The journey of Unison.gg—from concept to a functioning, decentralized ecosystem—offers a blueprint for what building in the Web 3.0 era really means. It's more than just smart contracts and tokenomics—it's about building systems of trust, fostering communities of governance, and designing financial architectures that are transparent, programmable, and inherently participatory.

Unison.gg is a testament to what happens when technology is leveraged not only to distribute value, but also to distribute control.

This chapter reflects on key themes from the book, synthesizes the lessons learned across product and finance, and offers a vision for the future of leadership in Web 3.0 ecosystems.

9.1 Key Lessons from the Unison.gg Journey

Lesson 1: Decentralization is not binary—it's a progression.

Unison.gg's evolution from a core team-controlled MVP to a DAO-governed protocol was gradual and intentional. It required building systems of accountability, incentivizing thoughtful participation, and ensuring redundancy in critical areas before handing over control.

Lesson 2: Product Managers Become Ecosystem Designers.

The PM in Web 3.0 is less of a sprint planner and more of a protocol architect. They curate incentives, translate governance into strategy, and coordinate across a decentralized network of stakeholders. In the absence of formal authority, influence is earned through clarity, credibility, and contribution.

Lesson 3: Finance Is Now Public Infrastructure.

Every transaction, payout, budget approval, and treasury allocation were visible on-chain. Finance became a participatory, data-driven system. DAO members didn't just track outcomes—they voted on them, proposed them, and sometimes contested them. This created unprecedented accountability, but also required financial fluency across the community.

Lesson 4: Community Is the Operating System.

Unison.gg scaled not through paid marketing but through creator guilds, regional ambassadors, and contributor networks. It has been proven that in Web 3.0, growth is not about acquiring users—it's about co-opting collaborators.

9.2 The Evolving Role of Product Leadership

The product leaders of tomorrow must be:

Technically fluent: Able to design with composability, protocol constraints, and chain-specific limitations in mind.

Economically literate: Capable of modeling token utility, staking incentives, and contributor payouts.

Governance-aware: Skilled in coordinating DAO workflows, managing voting fatigue, and translating community feedback into protocol design.

Globally minded: Comfortable leading pseudonymous teams, navigating asynchronous communication, and co-building with contributors in multiple jurisdictions.

PMs will also have to design for users who don't care about decentralization. They must simplify the user experience without compromising the fundamental principles of trustlessness and user ownership.

9.3 The Financial Stack of the Future

Finance in Web 3.0 will be defined by:

Treasury automation: Smart contracts will govern payments, vesting, budget approvals, and yield strategies.

On-chain accounting standards: Tools like Rotki, Coinshift, and Utopia Labs are setting new norms for transparent, compliant bookkeeping.

Composable capital: Treasuries will be deployed across lending, insurance, and LP positions with risk thresholds encoded directly into the protocol.

Stakeholder auditing: The financial health of an organization will be verifiable by token holders in real-time, shifting financial ops from internal gatekeepers to community stewards.

Finance will not be a back-office function—it will be a user-facing feature.

9.4 What Comes Next: Open Challenges

Despite progress, the path forward is not without obstacles:

Regulatory clarity: How do decentralized systems interface with increasingly strict KYC/AML requirements and national securities laws?

Governance legitimacy: What mechanisms can improve voter turnout, reduce manipulation, and ensure that token distribution aligns with protocol vision?

Security culture: How can teams prevent exploits without slowing innovation?

Inclusivity at scale: Can Web 3.0 reach beyond the crypto-savvy elite to empower creators and builders in underserved regions?

These are not just engineering questions—they are societal ones.

9.5 The Unchained Vision

Unison.gg was a living experiment. It demonstrated that:

- Creators can earn without platform gatekeepers.
- Contributors can work without employment contracts.
- Products can grow without centralized growth hacking.
- Organizations can operate without CEOs.

This is the unchained vision of Web 3.0:

- A world where platforms don't own users.
- Where code is law, and governance is collective.
- Where finance is transparent, and value is co-created.
- Where trust is built not through promises—but through verifiable execution.
- This book is both a record of that journey and a call to action.

To the next generation of builders:

- Let this be your invitation to design differently.
- To the next wave of investors:
- Let this serve as your framework for responsible support.

And to every user, creator, or community member, let this be your proof that the tools of ownership are already in your hands. The future of product and finance is not just decentralized. It is unchained.

Unchained Capital: Product Innovation and Financial Architecture in the Web 3.0 Era: How Unison Reimagined Startup Growth, Treasury Design, and Community-Led

Governance Overview

This part of the book examines how the rise of Web 3.0 is transforming the financial foundations of startups and digital platforms, from capital raising to ownership distribution. It examines token launches, community funding mechanisms, decentralized venture models, and hybrid legal and financial structures that bypass or reimagine traditional equity and venture capital frameworks.

By analyzing case studies, including DAOs, protocol-native treasuries, and emerging regulatory approaches, the book offers a practical and strategic guide for founders, investors, and contributors navigating the future of decentralized capital formation.

Chapter 1: The Death of Traditional Fundraising?

For decades, startup capital formation followed a centralized, investor-first model. Founders pitched venture capitalists, gave up equity for funding, and were guided—or pressured—toward high-growth exits. While this model enabled many successful tech companies, it systematically excluded global talent, misaligned incentives, and concentrated power in the hands of a few.

Web 3.0 challenges this paradigm. By introducing community-driven fundraising mechanisms—such as token offerings, DAOs, and decentralized liquidity pools—it reframes capital not just as a financial input, but also as a governance tool, an incentive layer, and a mechanism for user ownership.

This chapter explains why the traditional fundraising model is under strain and how Web 3.0 provides an alternative that is global, inclusive, and programmable.

1.1 A System Under Strain

Geographical Bias

The vast majority of venture capital flows to startups located in elite hubs such as Silicon Valley, New York, and London. Founders in LATAM, Africa, Southeast Asia, and Eastern Europe often face systemic barriers—not due to a lack of talent or ideas, but due to geographic disconnection from the traditional VC network.

Founder Dilution

Founders often lose between 15–30% of their company in early-stage rounds, long before product-market fit. With each new round, their stake decreases, reducing control and long-term upside. This dilution can stifle creativity and shift decision-making away from those closest to the product and users.

Exit-Driven Strategy

VCs invest with the expectation of a 10x+ return. This leads to a focus on growth-at-all-costs strategies and a rigid timeline for exits, typically 7–10 years. Founders may be pushed toward acquisitions or IPOs that serve investor liquidity but not necessarily product sustainability or user interests.

Closed Participation

Accredited investor laws restrict private equity participation to individuals who meet specific income or asset thresholds. As a result, users who generate platform value, such as early adopters, creators, and evangelists, are often locked out of early investment opportunities and wealth creation.

1.2 Web 3.0's Philosophical Shift

Permissionless Participation

Anyone with an internet connection and a crypto wallet can participate in a token sale, governance vote, or contributor bounty. There are no gatekeepers. Capital becomes accessible to students in India, coders in Nigeria, and creators in Argentina, on equal terms with institutional players.

Decentralized Allocation

Funding decisions are increasingly made by DAOs through public proposals and votes. Treasury management is enforced by smart contracts and multi-sig wallets—not centralized CFOs or boards.

Compositional Capital

A token is more than a fundraising tool—it's a programmable asset. It can enable payments, access control, staking rewards, governance voting, and more—all governed by code.

Incentive Alignment

Web 3.0 flips the model— users and contributors become co-owners. The more value you provide— through use, development, or governance—the more upside you gain. This transforms stakeholders into active network participants, rather than passive consumers.

1.3 The Rise of Protocol-Native Fundraising

ICO (Initial Coin Offering)

ICOs allowed teams to raise millions in capital via token sales with minimal friction. While revolutionary, early ICOs often lacked accountability and drew regulatory scrutiny for offering unregistered securities.

IDO (Initial DEX Offering)

IDOs shifted launches to decentralized exchanges. Liquidity is available instantly, but concerns remain regarding price volatility and front-running bots. Still, IDOs reduced gatekeeping and democratized access.

Fair Launches

No presale. No VC allocation. Tokens are distributed to early users via mining, staking, or participation. Yearn Finance famously launched this way, and the model has become a gold standard for decentralization-first communities.

Liquidity Bootstrapping Pools (LBPs)

LBPs allow token prices to start high and decay over time, deterring whales and encouraging broad participation. Utilized by projects such as Balancer and PrimeDAO, LBPs introduce fairness to launch mechanics.

1.4 Who's Raising Capital?

Protocols

Projects like Aave, Compound, and Arbitrum raise funds through governance tokens to reward usage, bootstrap liquidity, and fund development.

DAOs

Decentralized Autonomous Organizations, such as Gitcoin or Nouns DAO, raise funds to support grants, community initiatives, and infrastructure, all through community-led governance.

Creators & Collectives

Artists, musicians, educators, and organizers launch NFT collections, creator DAOs, and social tokens to directly fund projects with their audiences, eliminating the need for intermediaries.

Infrastructure Builders

Foundational teams building L2s, oracles, or dev tools raise tokens to fund their open-source ecosystems, often offering staking rewards or ecosystem incentives in return.

IV. Arbitrum Logo (Image Credits: Arbitrum)

1.5 Pitfalls and Risks

Speculation & Hype

Many early token launches were driven by hype rather than utility. Projects raised millions without a product, only to disappear or deliver little value. Speculation distorts project focus and user expectations.

Lack of Accountability

Anonymous teams with no public roadmap or delivery mechanisms led to a wave of "rug pulls." This eroded trust and attracted regulatory attention.

Regulatory Ambiguity

Tokens may be classified as securities, commodities, or utilities, depending on how they're issued and used. Lack of clarity exposes teams and investors to legal risk.

Governance Fatigue

Participating in decentralized governance is time-consuming. Without proper delegation, incentives, and tooling, voter turnout and proposal quality decline over time.

1.6 Hybrid Models: The Best of Both Worlds?

Token + Equity

Projects issue tokens for utility and community incentives, while issuing equity to institutional investors for legal clarity and long-term alignment.

DAO + Legal Wrapper

Protocols operate as DAOs but use a legal entity, like a foundation or DAO LLC, to interface with service providers, regulators, and fiat systems.

Convertible Token Warrants

Investors fund early development in exchange for future tokens, contingent upon the achievement of specific milestones. These contracts typically include vesting provisions, clawback clauses, and governance terms.

These hybrid models enable teams to maintain decentralization while upholding institutional legitimacy and compliance.

1.7 Conclusion

The traditional venture capital model isn't disappearing—but it is being challenged, reimagined, and expanded. Web 3.0 doesn't just change how capital is raised—it changes who gets to raise it, who gets to contribute, and who gets to benefit.

In the chapters ahead, we'll explore how token design, liquidity mechanics, treasury governance, and compliance tooling make decentralized capital not only possible, but sustainable.

Chapter 2: Token Launches and Capital Efficiency

In Web 3.0, tokens are more than just fundraising tools—they are programmable assets embedded with governance rights, incentives, and network effects. A token launch is not merely a capital event, but a strategic design mechanism that determines how value is distributed, who owns the protocol, and how aligned the stakeholders are.

This chapter examines the various models of token launches, their advantages over traditional capital formation, and how they can be designed to foster efficient, sustainable, and community-driven growth.

2.1 What Is a Token Launch?

A token launch refers to the initial public distribution of a blockchain-native token, enabling a project to raise capital, bootstrap liquidity, and initiate user participation. Unlike traditional equity, tokens serve multiple functions:

- **Utility**: Used to pay for services, unlock product features, or access gated communities.

- **Governance**: Provides holders with voting rights on protocol decisions, upgrades, and treasury allocations.

- **Incentive**: Rewards users, developers, and validators who support the network.

- **Store of Value**: Tokens can appreciate in value based on network growth and demand.

This multifunctionality makes token design inherently complex and strategically powerful.

2.2 Token Launch Models

1. Initial Coin Offering (ICO)

- **Description**: A smart contract accepts ETH or stablecoins in exchange for project tokens.

- **Strengths**: Direct, fast, borderless fundraising.

- **Risks**: Regulatory scrutiny, poor investor protections, many early ICOs failed or scammed users.

- **Example**: Ethereum raised $18 million in 2014 via an ICO to build its blockchain.

2. Initial DEX Offering (IDO)

- **Description**: Tokens are launched via automated market makers (AMMs) on decentralized exchanges.

- **Strengths**: Permissionless listing, real-time price discovery.

- **Risks**: Vulnerable to bots, slippage, and high volatility.

- **Example**: SushiSwap used an IDO to bootstrap its user base and fork Uniswap.

3. Fair Launch

- **Description**: All tokens are earned through usage, mining, or contribution. No pre-sales or allocations.

- **Strengths**: Strong community trust and decentralization.

- **Risks**: No up-front capital raised; potential for early dominance by insiders with better technical access.

- **Example**: Yearn Finance launched its YFI token with zero premine, earning trust and legitimacy.

4. Liquidity Bootstrapping Pools (LBP)

- **Description**: Tokens are sold via Balancer pools where the initial price is high and decreases over time.

- **Strengths**: Reduces front-running, encourages fairer distribution.

- **Risks**: Complex to execute; users need education to understand mechanics.

- **Example**: PrimeDAO used LBPs for its token launch to prevent early manipulation.

5. Bonding Curves

- **Description**: Token prices increase with demand according to a predefined formula.

- **Strengths**: Continuous fundraising, embedded pricing logic.

- **Risks**: High slippage at large volumes, susceptible to manipulation.

- **Example**: Aragon and Gnosis used bonding curves to issue governance tokens.

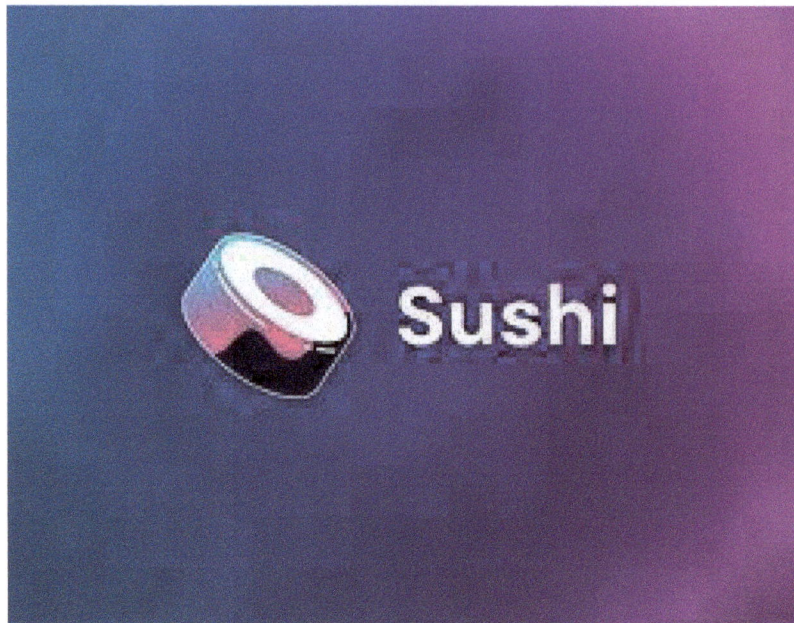

V. SushiSwap Logo (Image Credits: IQ.Wiki)

2.3 Token Launch vs. Traditional Equity Funding

1. Speed and Scale

A token launch can raise global capital in minutes through a smart contract, unlike the weeks or months needed for legal, negotiated equity rounds.

2. Cost Efficiency

Traditional fundraising includes lawyers, bankers, and regulatory filings. Token launches cost significantly less and scale via smart contract execution.

3. Incentive Alignment

Token distribution includes contributors, developers, and users, not just investors, resulting in better-aligned stakeholder incentives.

4. Composability

Tokens can interact with DeFi—be staked, traded, or used as collateral—to unlock liquidity and extend their utility.

5. Governance Integration

Launches can embed DAO voting from day one, establishing democratic control over treasury and protocol development.

2.4 Designing for Capital Efficiency

Key Mechanisms:

- **Token Allocation**: Prioritize users and contributors, and limit early investor allocations to prevent centralization of control.

- **Vesting Schedules**: Implement linear or cliff-based vesting for the team and investors to reduce short-term sell pressure.

- **Emissions Curve**: Control token supply over time to prevent inflation and sustain long-term value.

- **Governance Activation**: Encourage early token staking and delegation for meaningful participation in proposals.

Best Practices:

- Use **Token Engineering** simulations (e.g., BlockScience, TokenSpice) to model behavior.

- Deploy with trusted **smart contract libraries** such as OpenZeppelin.

- Consider **DAO-native launchpads** such as DAOhaus or Juicebox for community alignment.

- Integrate **real-time analytics dashboards** (e.g., Dune, Flipside Crypto) to track post-launch activity.

2.5 Risks and Mitigation Strategies

Risk	Mitigation
Price Volatility	Use LBPs or vesting schedules, limit the concentration of early holders
Regulatory Uncertainty	Use disclaimers, avoid promises of return, and use legal wrappers
Whale Dominance	Cap per-wallet contributions, implement quadratic voting
Poor Token Utility	Design real, protocol-native use cases for the token
Governance Apathy	Delegate roles, introduce voting rewards, and use off-chain tooling

2.6 Case Study: Optimism's Retroactive Airdrop

Strategy:

Optimism launched its token by rewarding past contributors and early users based on provable actions.

Impact:

- Created immediate goodwill and legitimacy.

- Avoided VC-heavy cap tables.

- Distributed governance power to the real builders of the ecosystem.

Lesson:

Token launches can reward value retrospectively, not just speculate on future potential.

2.7 Conclusion

Token launches are a foundational building block of decentralized capital formation. They combine financing, governance, and growth into a single event—when done right, they empower communities, align stakeholders, and unlock composable liquidity.

As the space matures, successful launches will be those that are designed with discipline, transparency, and token utility, not just hype. In the next chapter, we explore how communities themselves are becoming a source of capital, coordination, and innovation.

Chapter 3: Community as Capital

In traditional finance, capital is defined by monetary contributions—cash from investors, credit from institutions, or equity from shareholders. In Web 3.0, however, capital is multidimensional. Communities are no longer just users or supporters; they are builders, governors, marketers, and funders. Their contributions are often non-financial—measured in code, content, coordination, and culture—but they generate tremendous value and are increasingly compensated with tokens, influence, and ownership.

This chapter examines how community is emerging as the new source of capital in decentralized ecosystems, redefining who contributes value, how that value is measured, and how it is rewarded in Web 3.0-native projects.

3.1 Redefining Capital in the Web 3.0 Context

Traditional Capital = money, assets, equity.

Web 3.0 Capital includes:

- **Social Capital** – Community influence, reach, and credibility

- **Human Capital** – Developer time, creative work, and protocol contributions

- **Reputation Capital** – DAO governance activity, consistent voting, bounties completed

- **Cultural Capital** – Memes, onboarding materials, community norms

This distributed model empowers contributors from any background to add value and gain ownership. These new forms of capital, when properly structured and measured, can rival or even surpass traditional venture capital injections.

3.2 Community-Led Funding Models

1. Retroactive Public Goods Funding (RPGF)

- **Concept:** Projects are rewarded based on past contributions.

- **Example:** Optimism RPGF has distributed millions to builders, content creators, and infra contributors.

- **Why it works:** It rewards proof of work, avoids speculative funding, and aligns incentives around impact.

2. Bounty Boards

- **Platforms**: Gitcoin, Layer3, and Dework offer micro-tasks—docs, translations, code fixes.

- **Benefits**: On-demand task resolution, low cost, high engagement.

- **Risks**: Requires constant curation and contributor verification.

3. DAO Grants and Community Treasuries

- **DAOs like** Aave Grants, Uniswap Foundation, and Nouns DAO allocate millions via governance.

- **Structure**: Community voting or working group review ensures legitimacy and distributed oversight.

4. Creator DAOs and Social Tokens

- **Use Case**: Creators (artists, educators, developers) launch tokens to fund work and build aligned communities.

- **Examples**: Friends With Benefits (FWB), SongCamp, Forefront.

- **Impact**: Fans become investors; engagement becomes ownership.

3.3 Reputation-Based Systems and Contributor Frameworks

New infrastructure enables communities to reward non-monetary value through:

- **POAPs (Proof of Attendance Protocol)**: Attestation badges for participating in calls, events, or proposals.

- **Coordinape**: Peer-to-peer budgeting tool where contributors distribute tokens based on perceived value.

- **SourceCred**: Measures online contributions (GitHub, Discourse) and calculates a decentralized reputation score.

These systems enable DAOs to create inclusive, bottom-up economies with flexible labor structures and fluid ownership pathways.

3.4 Case Study: Gitcoin and Quadratic Funding

Gitcoin has distributed over $50 million to open-source builders, utilizing a community-first approach.

- **Mechanism:** Quadratic Funding (QF) multiplies small-dollar donations via a matching pool.

- **Outcome:** Projects with many small supporters receive more funding than those backed by a few large supporters.

- **Lesson:** Community alignment matters more than capital concentration.

This model democratizes funding and reflects the *community's will*, not just its wallets.

VI. Gitcoin (Image Credits: Phemex)

3.5 Community Labor as an Asset

In traditional firms, labor is contractual. In DAOs, labor is tokenized:

- **Work is compensated** in tokens that provide voting rights and economic upside.

- **Governance becomes compensation**—participants are rewarded for voting, proposing, and guiding.

- **Contribution → Ownership:** A contributor's labor can turn into significant equity-like upside in protocols that grow.

This fundamentally alters the employer-employee dynamic, as workers become co-owners of the protocol.

3.6 Challenges and Considerations

Challenge	Problem	Mitigation
Sybil Resistance	Fake identities and gaming funding	Use BrightID, Gitcoin Passport, Proof of Personhood
Governance Fatigue	Contributors disengage from endless proposals	Delegate systems, rotating stewards, and async voting
Contributor Discovery	High-value contributors overlooked	Reputation systems, peer nominations, and contribution logs
Treasury Mismanagement	DAOs overspend early funds or fund low-impact projects	Budget caps, vesting, milestone-based grants

Sustainable community capital requires operational maturity, not just enthusiasm.

3.7 Designing Community Capital Systems

- **Clear Onboarding Paths**: Guide users from passive members to active contributors.

- **Layered Incentives**: Combine financial rewards with governance power and social recognition.

- **Transparent Budgeting**: Use dashboards (e.g., Dune, Tally) to show fund flows and grant recipients.

- **Iterative Governance**: Regularly refine processes with retrospectives and proposal analysis.

The most successful DAOs treat their communities like both employees and investors, with clear expectations and shared rewards.

3.8 Conclusion

Communities in Web 3.0 are no longer peripheral—they are the infrastructure. Protocols grow, pivot, and govern themselves through community action. The labor, loyalty, and alignment of contributors are now the most scalable form of capital. And by distributing ownership through tokens, DAOs can convert culture into capital, without ever raising from traditional investors.

Chapter 4: Protocol Equity and Governance Tokens

In the traditional startup ecosystem, equity is the foundational mechanism for ownership, control, and value accrual. Founders, employees, and investors negotiate stakes in return for capital or labor, governed by a centralized board and shareholder agreements. But in Web 3.0, **tokens replace equity** as the primary unit of ownership. These tokens are programmable, liquid, and accessible globally, enabling a new model of **protocol-native ownership** through decentralized governance.

This chapter examines the emergence of governance tokens as a new form of equity, their distribution of power and value, and the structural and legal implications of this shift.

4.1 What Is a Governance Token?

A governance token is a blockchain-native asset that grants holders voting rights over decisions affecting a protocol's future. Unlike equity, which typically governs a private company via shareholder meetings, governance tokens operate in transparent, real-time, and code-enforced ways.

Key functions include:

- Voting on protocol upgrades and feature proposals

- Allocating treasury funds and grants

- Adjusting parameters (e.g., interest rates, fee splits, emissions)

- Electing delegates or working groups

- Signaling support for strategic direction

These tokens blend **control rights, utility value, and economic exposure**, effectively creating a public and programmable form of protocol equity.

4.2 Cap Tables vs. Token Distribution Models

In equity-backed startups:

- A cap table tracks share allocations among founders, investors, and employees.

- Equity is illiquid, with long vesting and limited transferability.

- Shareholder rights are contractual and jurisdictional.

In tokenized protocols:

- Distribution charts show token allocations across stakeholders (e.g., team, community, treasury, early backers).

- Tokens are liquid by design, with optional vesting and lockups.

- Rights are enforced via on-chain governance logic, not legal documents.

Common token allocation models:

- 25% to the community (via airdrops, incentives)

- 20–30% to core team (with vesting)

- 15–20% to investors (via token warrants or SAFTs)

- 15–30% to DAO treasury (for future use)

Token-based structures enable community members to become stakeholders from the outset, offering instant liquidity and participation.

4.3 Designing Governance Systems

Effective governance design strikes a balance between **participation, efficiency, and resilience.** Common structures include:

- **1 Token = 1 Vote:** Simple, but prone to whale dominance.

- **Delegated Voting:** Token holders assign voting power to active delegates, improving decision quality and turnout.

- **Quadratic Voting:** Vote weight increases at a diminishing rate to reduce whale influence.

- **Council-Based Governance:** Small elected teams handle day-to-day proposals, with community veto power.

Governance Layers:

- **Proposal:** Any token holder can suggest protocol changes.

- **Discussion:** Takes place on forums (e.g., Discourse, Commonwealth).

- **Snapshot Vote:** Off-chain signaling to test consensus.

- **On-Chain Execution:** Smart contracts enforce approved proposals.

4.4 Governance in Action: Notable Case Studies

Compound (COMP)

- The delegation model enables passive holders to assign voting power.

- Community decides interest rate models, market listings, treasury usage.

MakerDAO (MKR)

- DAO governance controls DAI's monetary policy, risk management, and operations.

- Uses Core Units and elected stewards to manage decentralized workflows.

Arbitrum DAO (ARB)

- Launched one of the largest token-based DAOs.

- Utilizes a Security Council, sub-DAOs, and on-chain grant allocations to scale governance.

4.5 Legal and Regulatory Considerations

Governance tokens blur the line between **utility and security**, raising regulatory scrutiny if they resemble investment contracts.

Risk factors include:

- **Expectation of Profit**: Token value appreciation tied to protocol success.

- **Centralized Managerial Effort**: Core teams running most operations.

- **Pre-product Sales**: Selling tokens before utility exists can resemble securities offerings.

Mitigation Strategies:

- Progressive decentralization: Gradually hand control to DAO participants

- Legal disclaimers: Clarify token purpose and avoid promises of profit

- DAO Wrappers: Form legal entities like Wyoming DAO LLCs or Cayman Foundations to shield contributors and interface with traditional systems

4.6 Economic Rights and Protocol Revenue

Not all governance tokens provide economic value, but some DAOs enable financial mechanisms like:

- **Staking Rewards:** Locking tokens to support protocol functions (e.g., validators, LPs) and earn yield

- **Revenue Sharing:** Distributing protocol fees or profits to token holders (e.g., SushiSwap, Curve)

- **Buybacks & Burns:** DAOs use revenue to reduce token supply, indirectly benefiting holders

Note: If designed improperly, these features may cause the token to be classified as a security under U.S. and global laws.

4.7 Governance Risks and Mitigations

Risk	Impact	Mitigation
Voter Apathy	Low participation undermines legitimacy	Vote incentives, quorum requirements, and delegation
Whale Dominance	Centralized control by large holders	Quadratic voting, wallet caps, staggered vesting
Capture or Collusion	Coordinated actors manipulate the treasury or proposals	Term limits, audit trails, transparency, open reviews
Smart Contract Exploits	Vulnerabilities in execution logic	Use of multi-sigs, formal audits, and phased governance

4.8 The Future of Protocol Equity

Governance tokens represent the **next evolution of corporate structure:**

- **Borderless:** No jurisdictional limitation on who can own or vote

- **Programmable:** Smart contracts enforce rules, unlock treasury, track KPIs

- **Dynamic:** Cap tables can evolve based on contribution, staking, or governance activity

Emerging trends:

- Cross-DAO governance interoperability

- Dynamic delegation models with reputation scoring

- DAO-native investment DAOs and treasury syndicates

- Legal hybrid models with off-chain arbitration or insurance mechanisms

Conclusion

Governance tokens are not just digital shares—they are tools for **co-ownership, coordination, and capital allocation** in decentralized ecosystems. As protocols shift from founder-led to community-governed, governance tokens will define who holds power, how value flows, and what kind of future these networks will build.

In the next chapter, we explore the **New Venture Stack**—how fundraising, investing, and equity mechanics are being re-engineered for the decentralized age.

Chapter 5: The New Venture Stack (Expanded)

Web 3.0 is more than just a new way to interact with the internet—it's a complete overhaul of how value is created, distributed, and owned. Nowhere is this transformation more evident than in venture funding. The New Venture Stack represents a shift from centralized, equity-based venture capital to decentralized, token-driven capital formation. This chapter explores each layer of this new architecture, its instruments, players, and how Web 3.0 startups finance growth without selling equity or sacrificing community alignment.

5.1 The Breakdown of the Old Stack

Traditional venture capital depends on:

- Gatekeeping: founders pitch, VCs decide.

- Time-boxed fundraising: staged rounds from Seed to Series D.

- Legal structuring: Delaware C-corps, legal opinions, shareholder rights.

- Exit outcomes: IPOs, acquisitions, secondary buyouts.

Problems:

- Excludes global and underrepresented builders.

- Constrains innovation with board control.

- Offers little to early users and contributors.

- Value accrues to a few, not the many.

Web 3.0 challenges this model with open access, fluid capital, and on-chain incentives.

5.2 Evolving Instruments of Capital Formation

SAFTs (Simple Agreements for Future Tokens)

- Legal contracts used in 2017–2021 boom.

- Tokens delivered post-network launch.

- Favored by U.S. legal counsel for Reg D compliance.

- Downsides: opaque, centralized investor access.

Token Warrants

- Conditional rights to tokens upon achieving defined milestones (e.g., mainnet launch).

- Used to synchronize VCs with long-term protocol growth.

- Often paired with equity or SAFTs.

IDOs (Initial DEX Offerings)

- Open token launches on DEXs like Uniswap or PancakeSwap.

- Fast, decentralized access to liquidity.

- Risk: high volatility, front-running bots.

LBPs (Liquidity Bootstrapping Pools)

- Launched by Balancer to improve fairness.

- Price starts high, then decays, reducing bot abuse.

- Promotes wide distribution over speed.

Airdrops

- Token distribution to users based on contribution or historical use.

- Aligns value with utility, encourages engagement.

- Often combined with governance onboarding.

Innovations:

- Streaming tokens via **Sablier, Superfluid**

- Soulbound tokens (non-transferable credentials)

- On-chain token vesting contracts replacing paper equity terms

5.3 Rise of Community-Led Venture Funding

Investment DAOs:

- **MetaCartel Ventures**: Operated by builders, funding dApps via DAO votes.

- **The LAO**: SEC-registered entity blending traditional legal protections with DAO-native mechanisms.

- **Orange DAO**: 1,200+ YC alumni pooling capital and deal flow.

- **Syndicate Protocol**: Tools for tokenized investment clubs and legal SPVs.

Benefits:

- Builder-first capital, faster execution, community curation.

Challenges:

- Legal ambiguity, herd bias in voting, governance fatigue.

5.4 Blending Equity and Tokens

Many projects raise both equity and tokens to access global capital while complying with legal structures.

Structure:

- Equity via SAFE/SAFT for VC investors.

- Tokens for governance, usage, and liquidity rewards.

Advantages:

- Access to institutional capital.

- Immediate liquidity and incentives for the community.

- Governance decentralization without regulatory overreach.

Tactics:

- Dual vesting schedules.

- Token warrant rights tied to equity investors.

- Parallel board rights and multi-sig participation.

5.5 DAO-Based Launch and Growth Infrastructure

Launch & Funding Platforms:

- **DAOhaus** – Deploy governance-ready DAOs using Moloch framework.

- **Juicebox** – Streaming treasury and token mechanics (used by ConstitutionDAO).

- **Questbook** – Grant tooling platform for DAOs.

- **Gitcoin** – Quadratic matching for public goods.

These replace traditional incubators and accelerators by decentralizing trust and treasury control.

5.6 DAO Treasury Design and Efficiency

Once capital is raised, DAOs manage it with precision and accountability.

Key Tools & Strategies:

- **Diversification** – Convert native tokens into stables, ETH, or yield assets.

- **Treasury Tools** – Enzyme, Karpatkey, Zodiac modules, Gnosis Safe.

- **Payroll & Streaming** – Sablier, Hedgey, Superfluid to compensate contributors.

- **Revenue** – Protocol-owned liquidity, yield farming, grants, service fees.

Operational Best Practices:

- Treasury dashboards

- Quarterly reporting to token holders

- Budget proposals and grant committees

DAOs become sovereign funds with operational autonomy and capital efficiency.

5.7 Legal Foundations of the New Stack

Legal wrapping is now modular and programmable.

DAO Legal Wrappers:

- **Wyoming DAO LLC** – Legal personhood for DAOs under U.S. law.

- **Swiss Foundations** – Common for protocol governance with a neutral jurisdiction.

- **Cayman Foundations** – Ideal for treasury management, used by dYdX, Arbitrum, and Uniswap.

Legal Tech Integration:

- **OpenLaw, LexDAO** – On-chain legal agreements and dispute resolution.

- **Aragon Court, Kleros** – Community-based arbitration services.

Web 3.0 legal systems are no longer just analog contracts—they're smart, composable, and interoperable.

5.8 Case Studies

Optimism

- Raised equity from Paradigm, a16z.

- Airdropped tokens to contributors and users.

- Allocated funding to Retroactive Public Goods Funding (RPGF).

Zora

- Received DAO-native funding from The LAO and Seed Club Ventures.

- Built open NFT infrastructure with creator-owned protocols.

Nouns DAO

- Self-funds via perpetual NFT auctions.

- Community decides treasury allocation for public goods, art, education.

These projects show how flexible, community-driven capital can outperform rigid venture structures.

Conclusion

The New Venture Stack is fluid, inclusive, and programmable. Capital flows through code, communities govern their treasuries, and builders can raise funding without begging gatekeepers.

This new paradigm unlocks:

- Global investor access

- Faster time-to-market

- Transparent treasury management

- Aligned contributor incentives

Chapter 6: DAO Treasury Architecture and Financial Governance (Expanded)

DAOs are financial organisms as much as they are governance experiments. In the absence of traditional CFOs and centralized finance teams, DAOs must rely on smart contracts, contributors, and automated tooling to manage their treasuries. Treasury architecture determines how well a DAO can survive market downturns, scale contributions, and fulfill its mission.

6.1 Why Treasury Architecture is Foundational

DAO treasuries:

- Fund public goods, contributor salaries, and infrastructure development

- Shape governance incentives and voting outcomes

- Represent protocol credibility to partners and regulators

A well-architected treasury is the heartbeat of protocol sustainability and community trust.

6.2 DAO Treasury Composition

Typical composition includes:

- **Native Governance Tokens** (e.g., UNI, COMP, ENS)

- **Protocol Revenue Assets** (e.g., fees in ETH, USDC)

- **Strategic Reserve Holdings** (blue-chip assets like ETH, BTC)

- **DeFi Positions** (LP tokens, staked assets, vaults)

- **NFTs or RWAs** (for creator DAOs or tokenized assets)

Diversification Strategies:

- Regular swaps of native tokens into stablecoins

- Controlled yield farming allocations

- Use of Protocol-Owned Liquidity (POL) to reduce dependence on mercenary capital

6.3 Treasury Infrastructure: Access and Control

Gnosis Safe is the default infrastructure for DAO treasuries, enabling:

- Multisig wallets with customizable signer thresholds

- Zodiac modules for enhanced role-based permissions

- Safe Snap integration to tie off-chain votes to on-chain execution

Advanced setups include:

- **SubDAOs with isolated budgets**

- **Oracles and automation** for payments and triggers

- **Streaming payments** via **Superfluid**, **Sablier**, or **Hedgey**

6.4 Treasury Operations: Compensation, Budgeting, and Payroll

DAOs must establish clear compensation and budgeting models:

Common practices:

- Salaries in stablecoins to minimize volatility

- Token grants with cliff and vesting

- Monthly or quarterly budgeting cycles

- Milestone-based grants and performance-based bonuses

Payroll Tools:

- **Sablier:** Time-based token vesting

- **Superfluid:** Real-time, continuous streaming

- **Parcel / Hedgey:** DAO-native HR dashboards and vesting control

Transparency is non-negotiable. DAOs often publish contributor addresses, rates, and deliverables in public forums or Notion hubs.

6.5 Treasury Analytics and Reporting

Tools for visibility:

- **Dune Analytics:** Custom SQL dashboards to track treasury flows

- **Flipside Crypto:** Event-driven financial analytics

- **Karpatkey Reports:** Institutional-grade reports for DAO clients

- **Internal metrics:** Runway, payment velocity, volatility exposure, spending cadence

Best practices include:

- Monthly/quarterly treasury updates

- Open-source dashboards and GitBooks

- Snapshot proposals tied to funding execution

6.6 Governance and Treasury Risk Management

Key financial risks:

- Token price collapse eroding runway

- Irregular or opaque contributor payments

- Treasury capture by small governance cliques

- Exploits via malicious proposals or misconfigured contracts

Mitigations:

- Multisig rotation and revocation

- Quorum thresholds and voting delays

- Audit trails, simulation testing, and timelocks

- Emergency pause contracts for treasury drains

Some DAOs simulate treasury scenarios using **TokenSpice**, **Gauntlet**, or custom frameworks.

6.7 Grant Programs and Ecosystem Incentives

Treasuries fund not just operations but ecosystem growth:

Types of Grants:

- **Open Grants:** Community proposals vetted by votes

- **Targeted Bounties:** For specific features or integrations

- **Quadratic Funding:** Gitcoin-style matching rounds

- **Retroactive Grants:** Rewards based on proven impact

Grants often account for 10–25% of treasury outflows, depending on maturity and mission.

6.8 Exemplars of Treasury Design

ENS DAO

- Three-segment treasury (Core Ops, Ecosystem, Endowment)

- ETH-denominated reserves with conservative budget proposals

- Transparent funding cycles and DAO forum oversight

Gitcoin DAO

- Treasury segmented into operational, public goods, and governance reserves

- RPGF used to reward impactful builders retroactively

- Custom grant tooling and open grant review pipelines

Uniswap Grants Program (UGP)

- Cross-functional review committee

- Budget limits and review timelines

- Tracking of grant milestones in public repositories

6.9 The Future of On-Chain Finance

The next evolution of DAO financial infrastructure includes:

- **AI-powered forecasting and budget simulations**

- **Autonomous treasury agents** for balancing reserves

- **Protocol-native taxation and royalty flows**

- **Financial NFTs** for streamed equity or grant rights

- **Cross-chain treasury coordination** with LayerZero, Axelar, or Wormhole

DAOs will eventually function like transparent sovereign wealth funds—programmable, autonomous, and embedded in code.

Conclusion

Treasury governance is the spine of decentralized finance. The more composable and transparent the system, the stronger the DAO becomes. With structured policies, resilient infrastructure, and automated oversight, DAOs transform from experimental collectives into sovereign financial institutions.

In the next chapter, we'll dive into how DAOs adapt under stress, whether through governance pivots, strategic reforms, or economic realignments.

Chapter 7: Adaptive Governance and Strategic Pivots in DAOs (Expanded)

Governance is not static. As DAOs mature, their structures often show signs of inefficiency, rigidity, or misalignment with evolving missions. Unlike corporations, where CEOs can pivot decisively, DAOs must evolve via decentralized consensus and encoded change mechanisms. Successful DAOs exhibit agility not through centralized leadership but by building resilient frameworks that allow for structured experimentation, stakeholder negotiation, and long-term cultural buy-in.

7.1 The Necessity of Adaptability in DAOs

A DAO is a digital institution governed by its community and constrained by its code. In such a system, failure to adapt manifests as:

- Protocol stagnation

- Governance apathy

- Capital misallocation

- Contributor burnout

- Competitive irrelevance

Adaptive DAOs recognize when to restructure, fork, or sunset obsolete components. They iterate governance like software—releasing v2 frameworks, modular upgrades, and leaner execution paths.

7.2 When to Pivot: Symptoms of DAO Inertia

DAO dysfunction often signals a deeper need for reform:

- Declining voter turnout and forum engagement

- Conflicting incentives between token holders, contributors, and users

- Redundant or bloated working groups with unclear KPIs

- Decision paralysis due to unclear leadership or broken quorum

- Treasury reserves are diminishing without measurable ecosystem growth

DAOs must track key performance indicators (KPIs) for governance health, including the turnout rate, forum participation, average time to proposal execution, and contributor retention.

7.3 Types of Strategic Pivots

1. Structural Pivots

- Transition from monolithic governance to modular "pods" or subDAOs

- Example: MakerDAO's Core Unit and Endgame architecture

2. Treasury Strategy Pivots

- Shift from spend-based to yield-generating strategies

- Diversification from native token reserves to stablecoins and staked assets

3. Incentive Model Recalibration

- Move from up-front grants to milestone-based or retroactive funding

- Reallocate rewards from governance mining to active contributors

4. Scope Realignment

- Exit non-core ventures and refocus on protocol primitives

- Spin-off of experimental projects into independent DAOs

7.4 Enabling Reform: Governance Design Mechanisms

- **Dynamic Quorums**: Adjust proposal thresholds based on engagement

- **Delegation Frameworks**: Empower informed and accountable voters

- **Term Limits & Rotations**: Refresh contributor and council roles

- **Sunset Clauses**: Automatically expire roles unless renewed

- **On-chain Feedback Surveys**: Gauge sentiment before final proposals

Resilient governance encodes its ability to adapt from the outset.

7.5 Tooling for Change

Tool	Purpose
Conviction Voting	Time-weighted support accumulation (used by 1Hive)
Ragequit	Exit with treasury share upon disagreement (Moloch)
SafeSnap	Execute off-chain Snapshot votes via on-chain automation
Karma, Agora	Track delegate performance, voting behavior, and reputation
Governance Routers	Plug-and-play governance modules (e.g., Compound Bravo)

These tools create **programmable pathways for reform** and emergency course correction.

7.6 Social Layers of Change

Hard governance rules often lag behind the soft power of community consensus. Informal influence matters:

- Discord moderators help enforce culture

- Delegates curate and frame agendas

- Governance podcasts, blogs, and AMAs inform and mobilize sentiment

Narrative power and community consensus are as critical as technical execution.

7.7 Case Studies

MakerDAO

- Introduced Core Units, then Endgame architecture for clarity and efficiency

- SubDAOs proposed to isolate risk and streamline decision-making

- Example of gradual but structured reform through proposal iteration

SushiDAO

- Survived leadership crisis by hiring a professional "Head Chef"

- Reallocated multisig controls, formalized treasury grants

- Community pushed back against opacity and revived contributor trust

Aragon DAO

- Abandoned court product and centralized Foundation

- Transferred treasury to DAO control and embraced open grants

- Demonstrates radical pivot from traditional nonprofit governance to DAO-first operation

Index Coop

- Refocused on product-level accountability

- Introduced working group budgets and contributor success metrics

- Regular retrospectives to refine governance and compensation

7.8 Resistance and Reform Fatigue

Governance change is hard:

- Stakeholders with power resist dilution

- Reform complexity creates voter fatigue

- Political infighting can paralyze action

Mitigation Strategies:

- Run simulations and test environments (e.g., GovSandbox)

- Use third-party governance coaches or facilitators

- Break reforms into digestible, modular proposals

- Include opt-in or phased deployment for controversial changes

7.9 Designing for Continuous Improvement

Best practices:

- Set governance KPIs and publish quarterly reports

- Fund governance research grants (e.g., via Metagov, BlockScience)

- Use open-source tools for real-time health monitoring (e.g., Agora, Tally dashboards)

- Conduct regular contributor and delegate feedback surveys

DAOs must institutionalize introspection and design governance like an evolving product.

Conclusion

Adaptability is not a bonus—it is a core feature of DAO longevity. Reform is not an admission of failure, but a celebration of resilience. DAOs that encode transparency, modularity, and iteration will outlast those rigid in form and slow in vision.

In the next chapter, we examine how transparency and financial disclosure enhance governance legitimacy, and how DAOs can set the standard for defining new benchmarks of public accountability.

Chapter 8: Financial Disclosure and DAO Transparency Standards (Expanded)

Transparency is the social contract of Web 3.0. In DAOs, where there is no central management, transparency becomes the mechanism through which legitimacy, trust, and accountability are achieved. DAOs don't just need transparency—they exist because of it. From open treasury dashboards to real-time contributor payment logs, DAOs are pioneering radical standards in financial visibility.

8.1 Why DAO Transparency Matters

In traditional organizations:

- Information is asymmetric and filtered through investor updates

- Treasury and compensation data are private

- Decision-making happens behind closed doors

In DAOs:

- Community members are stakeholders

- Governance relies on informed token holders

- Treasury funds are owned collectively

Transparency ensures:

- Contributors are compensated fairly and openly

- Voters know how funds are allocated

- Partners, developers, and users understand the protocol's health

8.2 Core Pillars of DAO Financial Transparency

Treasury Disclosure

- Real-time tracking of asset balances
- Liquidity pool positions, yield strategies, and diversification metrics
- Vesting schedules for token unlocks

Contributor Payment Visibility

- Wallet-level details of salaries, bounties, and grants
- Use of payroll tools like Parcel, Utopia, Hedgey
- Public compensation frameworks (e.g., role-based bands, milestone-linked payouts)

Proposal and Execution Reporting

- On-chain governance decisions via Snapshot and Tally
- Verified proposal execution scripts (e.g., ENS DAO, Nouns DAO)
- Delegation dashboards tracking voting history and rationale

Operational Metrics

- Contributor KPIs, grant effectiveness, treasury runway
- GitHub activity, retro reports, and forum engagement as impact indicators

8.3 DAO Transparency Infrastructure

Tool	Purpose
Dune Analytics	SQL-powered custom dashboards with real-time chain data
Flipside Crypto	Event-based on-chain analytics for performance and behavior tracking
OpenBooks	DAO-native accounting and ledger standard
Karpatkey	Treasury modeling and reporting for DAOs
Gnosis Safe	Secure multisig wallet with open transaction logs
Superfluid/Sablier	Streaming contributor payments with public visibility
Agora/Karma	Delegate behavior dashboards

Many DAOs also use GitBook, Notion, or GitHub to create human-readable financial libraries and documentation.

8.4 Case Studies in DAO Financial Disclosure

ENS DAO

- Treasury split into three spending categories: Core Ops, Ecosystem, Endowment
- Weekly contributor payments published via Google Sheets
- Proposals include public funding breakdowns and audit links

Gitcoin DAO

- Retroactive Public Goods Funding (RPGF) rounds with full payout data

- Budget reports visualized through Dune dashboards

- Steward Council publishes spending and decision retrospectives quarterly

Uniswap Grants Program

- All grantee results tracked on GitHub

- Community-facing Notion workspace includes proposal status and funds disbursed

- Voting records and summaries shared pre- and post-cycle

Nouns DAO

- 100% of treasury activity is on-chain and linked to verified proposals

- Every proposal's intent, execution script, and financial impact are published

- Uses daily auctions to fund the treasury, visible on public dashboards

8.5 Accounting Innovations in Web 3.0

DAOs require new frameworks for transparency:

- **DAO-GAAP**: A proposal for Generally Accepted Autonomous Principles

- **Proof-of-Spend**: Tie budget disbursements to on-chain proofs or KPI tokens

- **On-Chain Milestones**: Unlock contributor rewards only after verified outcomes

- **Subgraph-based ledgers**: Treat smart contracts as live accounting sources

- **Streaming Budgets**: Tokens released in real-time rather than lump sums

8.6 Barriers to Transparency

Barrier	Impact	Solution
Multichain Complexity	Assets fragmented across L1s, L2s, sidechains	Use aggregators like Zapper, Token Terminal
Contributor Privacy	Salary transparency may deter contributors	Use wallet aliases or role-based bands
Data Inaccessibility	Complex dashboards intimidate non-technical users	Create executive summaries and FAQs
Oversaturation	Too much raw data without interpretation	Curate visualized dashboards and retrospective reports

8.7 Best Practices for DAO Financial Disclosure

- Maintain a **monthly treasury snapshot** in both human and machine-readable formats

- Use real-time dashboards and link them in governance forums

- Publish **public-facing KPIs** for each team or grant

- Ensure **proposal execution data** is on-chain and auditable

- Create a **shared documentation space** for contributor roles and payment details

- Integrate **budget retrospectives** into every funding cycle

Transparency should feel like a feature, not a forensic exercise.

8.8 Designing Transparency into Governance

Embed transparency into operational design:

- Proposals must disclose the funding source and the spending plan

- Votes should be linked to deliverables or KPIs

- Grant recipients should upload GitHub logs, milestone reports, and expenditure receipts

- Reward transparency contributors (e.g., dashboard creators, data analysts) with DAO-native grants

A culture of transparency scales when **data literacy**, **tooling**, and **incentives** are aligned.

Conclusion

Financial disclosure is no longer a quarterly investor call—it's a real-time, participatory layer of decentralized trust. DAOs are at the forefront of redefining what openness, accountability, and fiscal governance entail. With the right tools, standards, and cultural practices, transparency becomes not only a legal shield but a competitive edge.

In the final chapter, we examine what this means for institutional trust and how DAOs may one day replace not only companies but also public infrastructure.

Chapter 9: The Future of Web 3.0 Finance and Institutional Trust (Expanded)

DAOs, tokenized treasuries, and community-led governance have introduced a new paradigm for building and maintaining trust. As centralized institutions falter under the weight of opaque processes and eroded public faith, Web 3.0 offers an alternative—one rooted in transparency, automation, and collective ownership.

9.1 From Institutional Hierarchies to Protocolized Trust

Traditional institutions depend on:

- Hierarchies and chains of command

- Regulatory enforcement and gatekeeping

- Audited reports and delayed accountability

Web 3.0 offers:

- Trustless execution via smart contracts

- Permissionless access to participation

- Real-time transparency and auditability

Protocols serve as rulebooks. DAOs serve as institutions. Blockchains serve as the court, ledger, and memory of record.

9.2 The Rise of DAOs as Autonomous Economies

DAOs represent:

- Micro-governments with constitutions and budgets

- Online communities with treasuries and workforces

- Global entities with no borders, time zones, or physical HQs

Future DAOs may:

- Issue community credit

- Fund infrastructure and education

- Maintain decentralized healthcare or insurance systems

- Reward citizenship and governance participation with real income streams

We are witnessing the emergence of **programmable societies.**

9.3 Public Goods and Protocol-Based Welfare

DAOs are uniquely positioned to fund and maintain digital public goods:

- Open-source software

- Developer ecosystems

- Educational content

- Civic infrastructure

New mechanisms include:

- **RPGF (Retroactive Public Goods Funding)**: Aligns capital with verified impact

- **Quadratic Funding**: Prevents plutocracy through matched donations

- **Streaming Universal Basic Income (UBI)**: Distributes payments in real time to contributors

In effect, protocols will not only pay developers, but they will also fund **digital civil services.**

9.4 DAO-to-DAO Economies

As composability increases, DAOs will:

- Collaborate via shared liquidity, treasury swaps, and cross-governance deals

- Form cartels, federations, or alliances with joint budgets and policies

- Trade access to user bases, tooling, and governance delegates

Examples:

- Protocol Guilds coordinating funding across ecosystems

- Gitcoin is integrating with Optimism, Arbitrum, and Base to distribute RPGF

- MetaCartel Ventures pooling DAO-native deal flow

These interactions form the foundation of **inter-protocol capital markets.**

9.5 Trust in the Age of AI and Automation

As AI agents and automated tooling enhance governance:

- Proposals can be ranked, summarized, and forecasted

- Delegates can be scored based on historical alignment and impact

- Risks can be modeled using on-chain behavior and treasury management metrics

This will transform DAOs into **autonomous fiduciaries.**

AI-enhanced governance will also allow:

- Delegation to algorithms based on community values

- Detection of sybil attacks or malicious governance patterns

- Simulation-based fiscal stress testing and budget proposals

9.6 Legal and Political Implications

As DAOs grow in influence:

- Governments will require disclosure, taxes, and consumer protection

- Jurisdictions may issue charters, licenses, or restrictions

- Hybrid entities (e.g., DAO LLCs, foundations) will become standard

In parallel, DAOs will:

- Form global governance frameworks

- Adopt legal wrappers to interact with the fiat world

- Build lobbying arms and self-regulatory frameworks

DAOs may evolve from **financial experiments into civic stakeholders.**

9.7 Long-Term Implications for Governance

Web 3.0 challenges us to rethink foundational questions:

- Who gets to decide how capital is allocated?

- What qualifies someone to govern?

- Can decentralized governance scale without becoming plutocratic?

- How do we build institutions that are inclusive, credible, and adaptive?

In this world, **governance is no longer a privilege—it's infrastructure.**

Conclusion

DAOs are building more than apps—they are building the foundations of a **new political economy.** One where participation is earned, not inherited. Where transparency is built-in, not retrofitted. Where institutions are fluid, composable, and collectively governed.

As we enter this new era of programmable trust, we must remember: **The DAO is not the destination—it's the vessel.** It's up to us to steer it toward a more equitable, transparent, and decentralized future.

Chapter 10: Augmenting Unison.gg with AI Agents – A Product Manager's Blueprint

In the rapidly evolving Web 3.0 landscape, artificial intelligence agents are emerging as powerful co-creators, capable of augmenting product, community, and financial operations at an unprecedented scale. For Unison.gg, the integration of autonomous and semi-autonomous AI agents marks a natural next step in decentralizing execution while enhancing real-time responsiveness.

This chapter outlines how a product manager can leverage AI agents to accelerate feature delivery, optimize DAO operations, and personalize user experiences within Unison.gg.

10.1 Why AI Agents in Web 3.0?

Web 3.0 has emphasized decentralization, transparency, and collective governance. However, these same principles often introduce operational overhead—slow proposal cycles, limited coordination bandwidth, and friction in content moderation or user support.

AI agents can:

- Operate 24/7 with autonomy and traceability.
- Interface with smart contracts, user data, and public chain analytics.
- Accelerate DAO governance cycles by summarizing proposals or clustering community sentiment.
- Power personalized fan engagement, creator insights, and automated treasury reporting.

For Unison.gg, where thousands of creators, fans, and token holders interact daily, AI agents offer scalable, protocol-aligned intelligence.

10.2 Use Cases: AI Agents Across Product Functions

1. Governance Optimization

- **Proposal Summarizer Agent:** Reads community proposals, summarizes key points, budget impact, and governance history for Snapshot/Tally voters.
- **Sentiment Clustering Agent:** Analyzes Discord/Discourse discussions to detect trending themes or voter fatigue signals.

2. Creator Support & Personalization

- **Creator Success Agent:** Provides real-time onboarding, dashboard walkthroughs, and best practice prompts based on creator profile and engagement history.
- **NFT Metadata Optimizer:** Suggests metadata improvements based on creator category, fan behavior, and historical NFT performance.

3. Treasury and Financial Automation

- **Treasury Monitor Agent:** Tracks burn rates, stablecoin reserves, and token volatility; alerts on budget anomalies.
- **Grant Impact Evaluator:** Utilizes on-chain metrics to evaluate the outcomes of funded proposals and generate retroactive grant scorecards.

4. Product Analytics & Iteration

- **Roadmap Feedback Synthesizer:** Clusters and summarizes community comments on roadmap Notion docs.
- **Experimentation Agent:** Proposes A/B test cohorts based on user segmentation and deploys personalized UX variants using zkAuth-compatible credentials.

5. Community Growth & Moderation

- **Language Localization Agent:** Translates platform content and DAO updates in real-time to support global contributors.
- **Spam Filter Agent:** Detects sybil behavior, raid attempts, or wallet spam in Discord, Snapshot, and proposal comments.

10.3 Agent Stack: From LangChain to On-Chain Execution

Unison.gg's AI agent system could be architected using:

- **LangChain:** For chaining tools and prompt-augmented reasoning.
- **LangGraph:** To model inter-agent workflows like sentiment summarization → proposal drafting → delegate notification.
- **CrewAI:** To coordinate agents by roles (e.g., Financial Analyst, Content Reviewer, Governance Steward).
- **IPFS/Filecoin:** For off-chain secure storage of chat logs, analytics, or agent observations.
- **Safe/Gnosis Modules:** For multisig-controlled execution of agent-recommended disbursements or alerts.

Example:

A "Proposal Insight Agent" (built on LangChain) summarizes a governance proposal → Sends it to "Delegate Sync Agent" (CrewAI node) → Publishes a summary with vote recommendation into Snapshot UI → Logs impact on IPFS.

10.4 Human-in-the-Loop Design

Despite their autonomy, AI agents must operate within a human-centric and decentralized ethical framework.

- All agent outputs should be transparent, signed, and auditable.
- Delegates and contributors can override, approve, or fine-tune agent recommendations.
- High-risk functions (e.g., token transfers, proposal submissions) require human multi-sig co-signature.

This preserves DAO agency while maximizing agent efficiency.

10.5 KPIs & Monitoring

Success metrics for AI agents should include:

- Proposal engagement rate improvements
- Time-to-decision reduction in governance
- Creator onboarding drop-off reduction
- Treasury anomaly detection lead time
- Cost savings in moderation and support

Conclusion

AI agents are not a threat to decentralization—they are its accelerant. At Unison.gg, they can help product managers act as orchestrators of machine-human cooperation, where code, community, and cognition interweave. Used ethically and transparently, these agents can scale governance, as well as precision and personalization. The next frontier is not automation alone—it is "autonomous collaboration" for a decentralized creative economy.

Chapter 11: Infusing GenAI into Unison.gg – Ethics, Guardrails & Unbreakable Security

As Unison.gg scales, integrating Generative AI (GenAI) into its product ecosystem presents both opportunities and responsibilities. This chapter examines how a Product Manager can lead the Responsible adoption of GenAI by prioritizing trust, compliance, and system-level resilience in Web 3.0 applications.

11.1 Expanding the Product Offering with GenAI

- **Content Generation Agents:** Generate proposal drafts, roadmap language, and release notes based on DAO inputs.
- **Visual Creators:** Empower NFT creators with GenAI-powered image generation tuned to style tokens.
- **Synthetic Community Training Bots:** Use fine-tuned LLMs on Unison.gg Discord logs to simulate user types for testing.

11.2 Guardrails and Ethical Considerations

Bias Minimization: Regular auditing of training data and model prompts to prevent social, economic, or regional biases.

Prompt Injection Protection: All input to GenAI modules must be filtered and sanitized using robust security measures.

Consent-Aware Learning: If models are exposed to user-generated content (UGC), explicit user consent must be logged via zkAuth or on-chain opt-ins.

11.3 Preventing Cyberfraud in a Web 3.0 Context

AI Wallet Watchdog Agents: Monitor multi-sig wallets and treasury contracts for anomaly-based access patterns using behavior-based ML models.

Phishing Detection Layers: Scan GenAI-generated messages for malicious intent before posting in public DAO or Discord spaces.

Zero-Knowledge Proof Validations: Validate sensitive user responses or signature approvals through zkSNARK verifiers where privacy is essential.

11.4 Designing Uncompromisable AI Infrastructure

- All GenAI services must be:
- Encrypted at rest and in transit using TLS 1.3+
- Logged immutably on IPFS with signed hashes
- Governed by an Ethics Multi-sig that can pause/review model behavior
- Deploy using confidential computing nodes (e.g., Azure DCsv3, GCP Shielded VMs) to ensure inference secrecy.

Conclusion

GenAI is a force multiplier—but only when rooted in transparency, restraint, and resilience. At Unison.gg, the goal is not just to innovate faster, but to innovate responsibly, respecting the autonomy of users, creators, and code. Through careful architecture and thoughtful governance, GenAI can be the foundation for a safer, smarter, more equitable Web 3.0 product ecosystem.

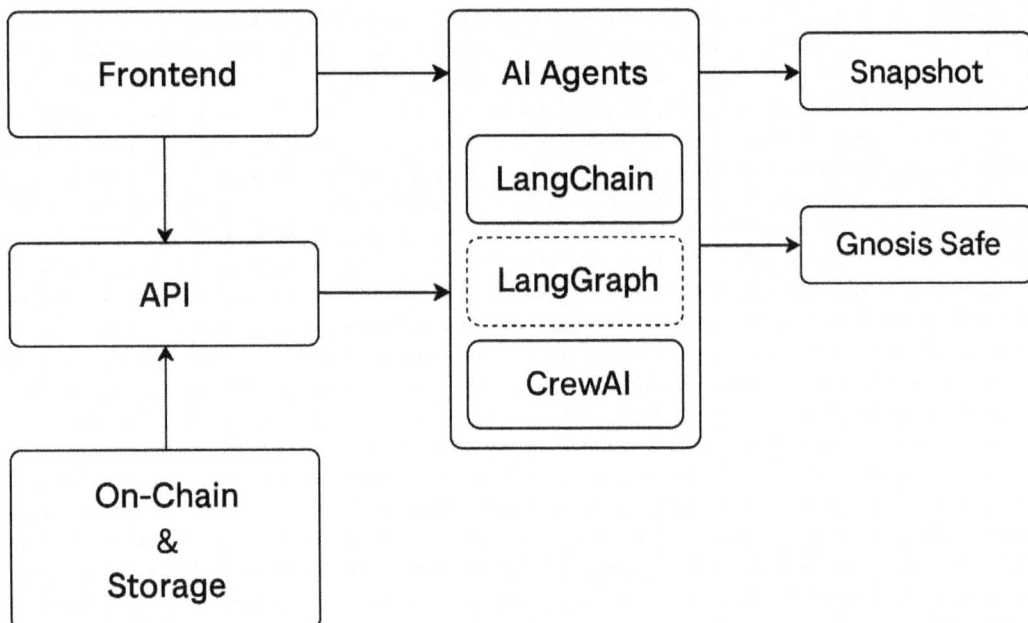

Chapter 12: Strategic Pivot to Bespoke Vault Architecture for Institutional Engagement

In its early phase, Unison.gg served as a comprehensive Web3 platform, integrating creator engagement tools, community governance layers, and token-based incentive mechanisms. However, after observing market signals and usage patterns, the team executed a strategic pivot: transitioning toward a bespoke vault infrastructure model specifically tailored for institutional investors and asset managers.

12.1 Why the Pivot?

Investor-Led Demand: Institutions sought programmable vaults with DAO-level oversight, yet without full exposure to community governance friction.

Custom Compliance Needs: Regulated asset managers required configurable rule sets—something general-purpose DAO tooling couldn't support.

Scalability & Capital Efficiency: Vault-based architecture allowed better pooling, delegation, and performance-based disbursement.

12.2 Features of the Bespoke Vault Model

Composable Vault Templates: Clients can choose templates (e.g., grants vault, yield split vault, NFT reserve vault) with customizable parameters.

On-Chain Governance Hooks: Vaults optionally plug into Snapshot or Safe module workflows.

AI-Aided Allocation: AI agents can co-simulate vault deployment scenarios, optimize funding flows, and flag misalignments before execution.

Auditable Impact Frameworks: Each vault includes post-facto impact measurement built on-chain.

12.3 Value Proposition for Institutions

Security: Gnosis Safe-compatible, battle-tested modules

Compliance: KYC-linked zkAuth layers, audit trails

Customizations: Vault-level logic authored via simple YAML-like DSLs, AI-assisted if desired

Trustless Transparency: IPFS-anchored vault logs and proofs of governance

Conclusion

This strategic pivot aligns Unison.gg not just with the ethos of decentralization, but also with the economic engine of capital allocators. By abstracting governance and rearchitecting for vault-first deployment, Unison is now positioned as a modular coordination layer—ready for sovereign funds, philanthropic DAOs, ecosystem treasuries, and hybrid investment syndicates.

Chapter 13: Unison.gg as a Groundbreaking Frontier in Web 3.0 and Institutional Finance

As Web 3.0 matures from idealism to infrastructure, few platforms reflect this transition more compellingly than Unison.gg. With its convergence of decentralized governance, AI integration, compliance-centric vault architecture, and product agility, Unison is no longer just a Web3 tool—it is a financial and coordination substrate for a new generation of investors, builders, and institutional stewards.

13.1 A Platform Engineered for the Next Wave of Capital

Unison.gg offers institutional investors and venture capitalists a rare blend:

Customizability: Vaults and modules are fully tailorable, allowing for customization down to the level of compliance logic and allocation granularity.

Interoperability: Integrates natively with Ethereum-compatible ecosystems, Snapshot governance, and zk-authenticated workflows.

Intelligence by Design: Native support for LangChain-, CrewAI-, and GenAI-based agent execution and insights.

This makes Unison.gg not just a platform but an adaptive infrastructure ready to scale from DAO tooling to sovereign-grade asset coordination.

13.2 Unlocking New Paradigms in Institutional Crypto Onboarding

Unison empowers institutional players to:

Run programmable treasuries with real-time AI anomaly detection

Deploy mission-driven capital via grant vaults, R&D pods, or impact-aligned distributions

Test on-chain strategies safely, with simulation environments, retroactive scoring, and vault-level governance circuit breakers

Integrate cross-border contributors securely with verifiable credentials and multi-language UX

13.3 Why Venture Capital Should Take Notice

Venture funds looking to invest in infrastructure, creator economies, or financial coordination tools will find in Unison:

A modular protocol with recurring revenue potential (vault-as-a-service)

A deep moats architecture through ethics-encoded GenAI layers and real-time governance intelligence

A community-aware roadmap that blends bottom-up innovation with product discipline

13.4 Forward Outlook

Unison.gg's strength lies in its ability to balance three pillars:

Trustlessness, through blockchain-native coordination

Adaptability, through composable vault and agent stacks

Ethical intelligence, through accountable GenAI infrastructure

In the years to come, Unison could evolve into a foundational protocol for institutional DAO rails, powering syndicates, treasury orchestration, and collaborative R&D ecosystems at a global scale.

For venture capitalists and strategic investors, this isn't just a platform. It's an onramp to the next layer of programmable capital.

About the Author

Dr. Srinidhi Vasan

VII. Dr Srinidhi Vasan (Image Credits: Forbes)

Dr. Srinidhi Vasan, CAPM, is a leading expert in financial innovation, with a focus on fintech, ESG-driven investment strategies, and digital payment systems. As the founder of Viche Financials, Dr. Vasan designs advanced financial frameworks that merge emerging technologies with sustainability goals to drive both economic and environmental impact.

Holding a Doctorate in Business Administration from Manipal GlobalNXT University and a master's in finance from Hult International Business School, Dr. Vasan brings strong academic and analytical expertise to their work. Their research, published in high-impact journals, spans AI-enhanced payment systems, blockchain applications, fraud detection, and sustainable investment metrics.

A recognized thought leader and reviewer in cyber-physical systems and ESG compliance, Dr. Vasan also serves as a Rotary International Ambassador, driving strategic initiatives across complex, multi-stakeholder environments. Their work continues to shape the future of finance through a blend of innovation, rigor, and real-world impact.

Mr. Sudarshan Chandrashekar

VIII. Mr. Chandrashekar (Image Credits: Merit Line)

Mr. Chandrashekar is a seasoned technical architect, author, and inventor with deep expertise in product development and financial technology innovation. He currently serves as Senior Technical Architect at DataCaliper Inc. and Chief Product Officer at a Web 3.0 cross-chain investment startup, where he has redefined product workflows to compete with industry leaders like Yearn Finance. In this role, he has led feature enhancements, improved user experience, and raised millions in seed funding, while working closely with financial institutions and retail investors.

With a background that includes consulting roles for top-tier banks such as Goldman Sachs and Wells Fargo, as well as billion-dollar blockchain startups, Mr. Chandrashekar has consistently delivered transformative solutions in fintech and enterprise cloud migration—most recently for a major Dallas-based airline.

As an inventor, he is awaiting patent approval for Auto Revive, a retrofit safety device designed to help vehicles float during flash floods, currently under review by American Honda Motor Co. He is also developing a Multi-Agent AI Copilot system to support the full lifecycle of automotive design and development.

Mr. Chandrashekar holds a bachelor's degree in telecommunications engineering from PES Institute of Technology, a Master of Science in chip design from Manipal Institute of Technology, and additional U.S.-based credentials. He continues to engage in lifelong learning through executive education at Harvard Business School. His contributions have earned him awards from V2 Technologies and Ikcon Technologies for excellence in cloud innovation and technology leadership.

www.ingramcontent.com/pod-product-compliance
Lightning Source LLC
Chambersburg PA
CBHW052344210326
41597CB00037B/6246